THE POLLINATION OF FRUIT CROPS

NB

Northern Bee Books
Scout Bottom Farm
Mytholmroyd
Hebden Bridge
HX7 5JS (UK)

www.northernbeebooks.co.uk
Tel: 01422 882751

ISBN: 978-1-912271-18-4

This title is reprinted by Northern Bee Books from the original first published in this form by The Horticultural Educational Association in 1961. The HEA had reprinted it from articles previously published in Scientific Horticulture Vols. XIV & XV

Printed by Lightning Source, UK

THE POLLINATION OF FRUIT CROPS

prepared by

The Fruit Committee

of

THE HORTICULTURAL EDUCATION ASSOCIATION

and reprinted from

Scientific Horticulture Vols. XIV and XV

NB

CONTENTS

THE POLLINATION OF FRUIT CROPS

Pollination is the most critical stage in the production of a fruit crop depending, as it does in most cases, on the activities of insects. The first link in the chain of events leading to fruit set is the transference of a pollen grain to the stigma of the flower. Successful pollination and the fruit set which follows are influenced directly and indirectly by many factors, so that an understanding of this process is important for all who are concerned with the cultivation of fruit.

In practice it is difficult to separate pollination from the events which immediately follow it—pollen germination, tube growth, fertilization, seed development and fruit drop. All these aspects have therefore been included in this review, which has been prepared to meet the needs of fruit growers, advisory officers, teachers and students. Part 1 deals with the basic principles of pollination and fruit-setting and the way in which these processes are influenced by internal and external factors. Part 2* describes the application of this knowledge to the practical problems of fruit growing in the British Isles.

ACKNOWLEDGEMENTS

This memorandum has been prepared by members of the Fruit Committee and the Association thanks the members concerned for their valuable contributions.

The members of the sub-committee responsible for the preparation of this publication were Mr. N. Stewart (Chairman until March, 1959), Dr. L. C. Luckwill (Chairman since March, 1959), Mr. A. G. Healey (Secretary until May, 1956), Mr. D. W. Way (Secretary since May, 1956), Mr. J. B. Duggan, Mr. A. Gavin Brown, Mr. W. G. Kent, Dr. W. S. Rogers.

This sub-committee acknowledge the help of many people who have assisted them by giving advice and by making available unpublished research results. In particular they wish to thank Mr. G. E. Clothier, Mr. R. M. Fulford, Miss E. M. Glenn, Dr. D. W. P. Greenham, Mr. W. H. Hogg, Dr. I. M. Modlibowska, Mr. A. P. Preston, Mr. G. C. White, Mr. R. R. Williams and Dr. D. Wilson. To all these people the Association extends its grateful thanks.

The second and third parts of the memorandum, the first part of which appeared in *Scientific Horticulture*, Vol. XIV, have, like the first part, been prepared by members of the Fruit Committee and the Association thanks the members for this second valuable contribution.

The members of the sub-committee responsible for the preparation of these parts were Dr. L. C. Luckwill (Chairman), Mr. D. W. Way (Secretary), Mr. J. B. Duggan, Mr. A. G. Healey, and Dr. W. S. Rogers. The preparation of the contributions and the compilation of Part I was undertaken by Dr. L. C. Luckwill, Mr. D. W. Way was the author of Part II, and Mr. Duggan undertook the task of compiling and checking Part III.

ACKNOWLEDGEMENTS

The Fruit Committee of the Horticultural Education Association gratefully acknowledge the help of many who, by making available unpublished data or by virtue of their advice, have assisted in the production of these final parts of "The Pollination of Fruit Crops". In particular they wish to thank Mr. A. Gavin Brown, Dr. C. G. Butler, Mr. B. A. Cooper, Miss E. M. Glenn, Mr. E. G. Gilbert, Mr. J. M. S. Potter, Miss A. Rake, Mr. J. H. Walker and Mr. R. R. Williams. The Committee is also indebted to a number of commercial fruit growers and others who have given information on various points.

The Council of the Association is greatly indebted to all these people for the considerable amount of work involved in this task.

PART I

FACTORS AFFECTING POLLINATION AND FRUIT-SETTING

The development of a flower into a fruit is known to growers as *fruit-set*, a term that is often used in a quantitative way to indicate the proportion of flowers that so develop. In the tree fruits the proportion of flowers which show clearly visible signs of developing into fruitlets within two weeks of bloom is known as the *initial-set*, whilst the term *final-set* refers to those fruits which survive to maturity.

Before describing the processes leading to fruit-set and the way they are influenced by internal and external factors, the structure and functions of the flower, and the behaviour of the chromosomes must be made clear.

Flower Structure

The structure of the simple flower of the plum is illustrated diagramatically in Text Fig. 16A. The male organs are the stamens, each consisting of a filament bearing an anther with four pollen sacs. When fully developed the anthers dehisce, exposing the pollen grains in which the male germ cells are later formed. The pistil, or female part of the flower, consists of the stigma and style, and the ovary which, in the plum, contains two ovules (Text Fig. 16A). Within each ovule is found an exceptionally large cell—the embryo sac—which at maturity contains several small cells, one of which is the female egg cell (Text Fig. 16c).

The Living Cell and the Nucleus

Every living cell of the plant contains a roundish body of denser structure than the rest of the protoplasm. This is the nucleus and, just prior to cell division, it can be seen under the microscope to contain a number of rod-like bodies—the *chromosomes*. The number of chromosomes per cell is a constant and characteristic feature of each species. Most plants are diploid, which means that they contain two complete sets of chromosomes in each cell. Thus the apple contains two sets of 17=34, and the black currant two sets of 8=16. When the cell divides the chromosomes also divide by splitting longitudinally so that each daughter cell receives an identical complement.

During the formation of the male and female sex cells another type of division occurs. Instead of each chromosome splitting in half, the two sets of chromosomes in the cell separate from one another, so that each daughter cell receives one complete set. As a result of this *reduction division* each pollen grain and egg cell contains only half the number of chromosomes present in the body cells of the plant. At fertilization, when the male and female sex cells fuse, the two sets of chromosomes come together again to restore the diploid complement.

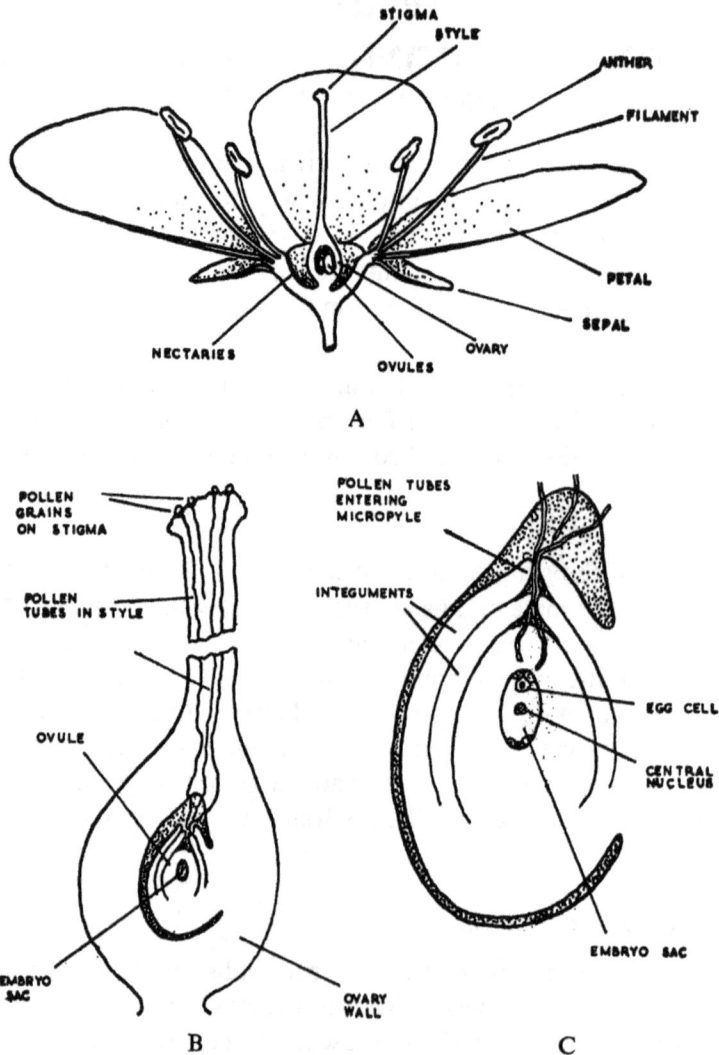

TEXT FIG. 16.

A—Semi-diagrammatic section of a plum flower. Some of the stamens have been removed
to clarify the diagram.
B—Longitudinal section through the pistil of a plum flower.
C—Longitudinal section through the ovule of a plum.

The great significance of the chromosomes lies in their function as carriers
of the hereditary material—the *genes*—which are located at definite points
along the length of the chromosomes. Each chromosome may carry hundreds
of genes, each of which determines some specific structural or physiological
character of the plant, and it is through the genes that these characters are
transmitted to the next generation.

The characteristic variability of plants, and indeed of all living things, is
due to the fact that the genes are unstable bodies which, either spontaneously
or under the influence of external agencies, may undergo slight changes or

mutations, each of which is reflected in some corresponding change in the organism. As a result of these mutations each gene normally exists in a number of different forms or *alleles*, any one of which may be present on a particular chromosome. Appreciation of this point is of particular importance for an understanding of the genetics of incompatibility (p. 137).

Pollination and Fertilization

When the flower is fully expanded the anthers dehisce and the stigma becomes receptive by secreting a sugary substance essential for the germination of the pollen grains. The flower is then ready for pollination. If a pollen grain of the right kind reaches the stigma it germinates, producing a long pollen tu'e which penetrates the stigma, grows down the style and enters the ovule, usually through a small pore known as the micropyle, where it bursts, liberating two male nuclei. One of these nuclei fuses with the egg cell whilst the other fuses with the large central nucleus of the embryo sac. This fusion of nuclei constitutes the process of fertilization. The fertilized ovule develops into a seed, consisting of an embryo, derived from the fertilized egg cell, endosperm tissue develops from the fertilized central nucleus, and the seed coat is formed from the outer integument layers of the ovule (Text Fig. 17). The endosperm is important as a source of nourishment for the embryo during its development. Hormones that control the growth of the embryo and the growth and abscission of the fruit are also secreted by the endosperm.

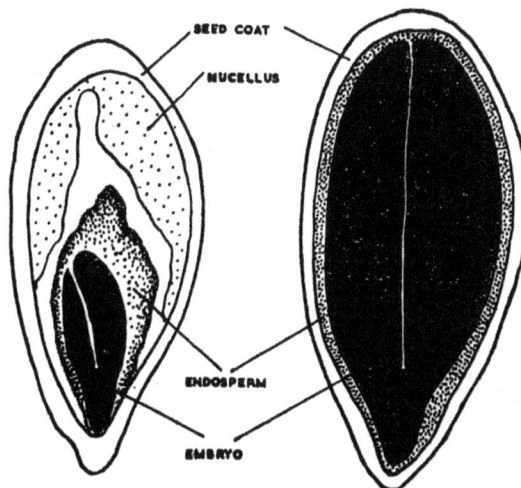

SEED COAT
NUCELLUS
ENDOSPERM
EMBRYO

TEXT FIG. 17. A PARTIALLY DEVELOPED (LEFT) AND MATURE (RIGHT) APPLE SEED IN LONGITUDINAL SECTION.

The Importance of Seeds for Fruit Development

The development of the ovary and surrounding tissue into a fruit is dependent on hormonal stimuli derived chiefly from the developing seeds (Luckwill, 1949) and, with few exceptions, fertilization and seed development are essential for fruit production. The influence of the seed on the growth of the fruit can be seen quite distinctly in peaches, which like all stone fruits, have an ovary containing two ovules. Following fertilization it is normal for one ovule to develop at the expense of the other, leading to a one seeded stone. As a result the fruit develops asymmetrically (Text Fig. 18A), the side on which the growing seed is situated being slightly larger than the other (Tukey, 1936).

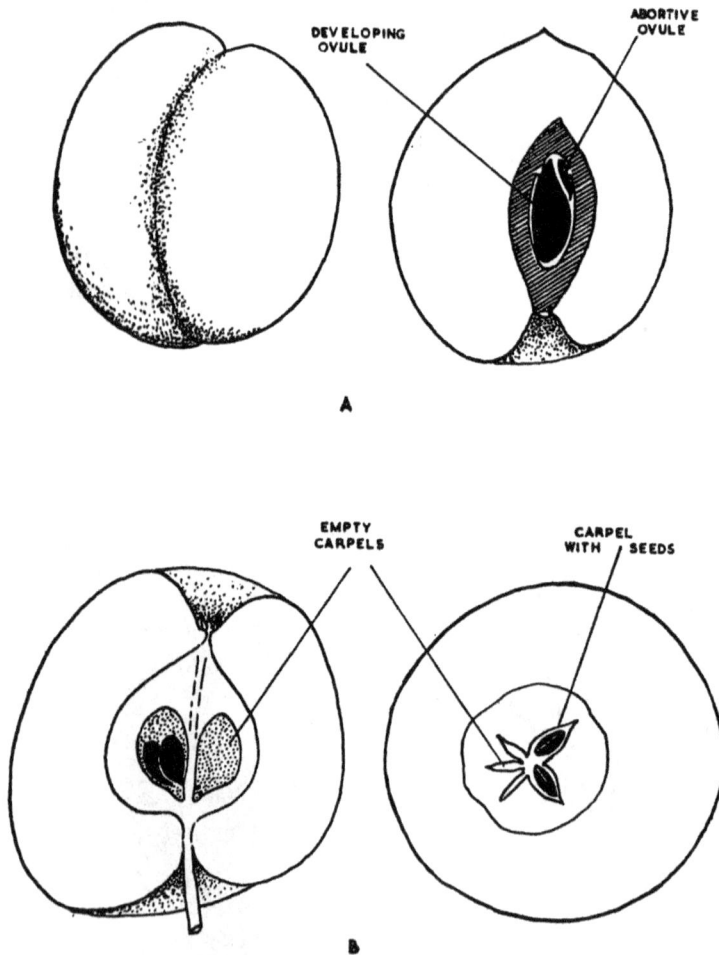

TEXT FIG. 18.

A—Asymmetrical growth of the peach in relation to ovule development (after Tukey, 1936).
B—Unequal development of opposite sides of an apple due to imperfect pollination.

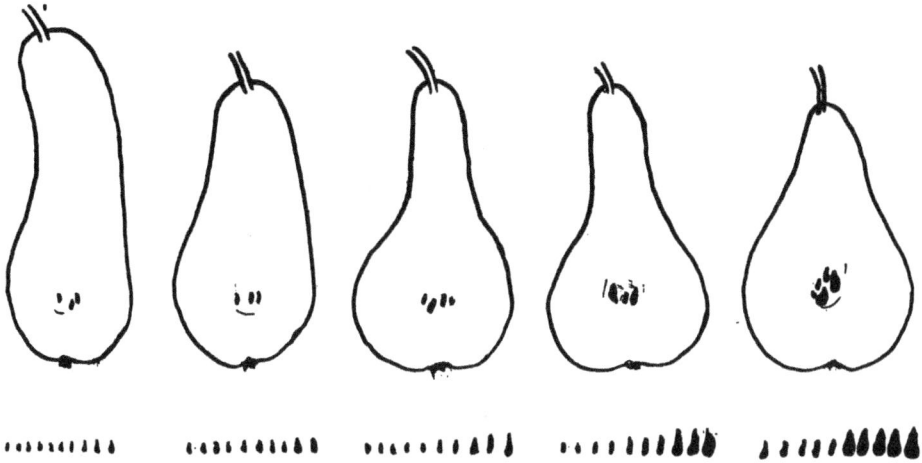

TEXT FIG. 19. THE RELATIONSHIP BETWEEN THE SHAPE OF CONFERENCE PEARS (ABOVE) AND THEIR SEED CONTENTS (BELOW). (AFTER SCHANDER, 1955B).

The flowers of apple and pear differ from those of the stone fruits in having five pistils, each containing two ovules. Although fertilization is essential for fruit development, it is not necessary for all ten ovules to develop and fruits may be produced with only a few seeds. In some varieties of apple such as Early Victoria and Grenadier, and in pears (Text Fig. 19), fruits with low seed contents are more elongated in shape than those having the full complement of ten seeds. In other varieties low seed numbers are associated with the development of lop-sided fruits (Text Fig. 18B), owing to the cortex of the fruit being stimulated to a greater degree in the region of those ovaries containing fertilized ovules (Schander, 1955a). In other varieties seed content seems to have no effect on fruit shape. In the pome fruits not only the shape but also the size of the fruit may be influenced, among other factors, by seed number. The correlation is usually positive, but under certain conditions where water or nutrients are limiting, there may be competition between the growing flesh of the fruit and the seeds, so that fruits with high seed numbers may actually make less growth than those containing somewhat fewer seeds (Schander, 1956).

In cane fruits and the strawberry the situation is a little different, for these fruits are compound, each flower containing a large number of ovaries (Text Fig. 20A), each with a single ovule, practically all of which need to be fertilized for normal fruit development. In these fruits lack of complete pollination and fertilization leads to malformation of the berry (Nitsch, 1952). Such malformation is often marked in strawberries (Text Fig. 20B) in seasons when weather conditions have been unfavourable for pollination or when partially male-sterile varieties have been planted without a suitable pollinator.

Causes of Fruit Drop in Apples

Flowers in which fertilization of one or more ovules has taken place soon start to swell and develop into small fruitlets, but in the tree fruits only a small proportion of these normally develop to maturity, the remainder being

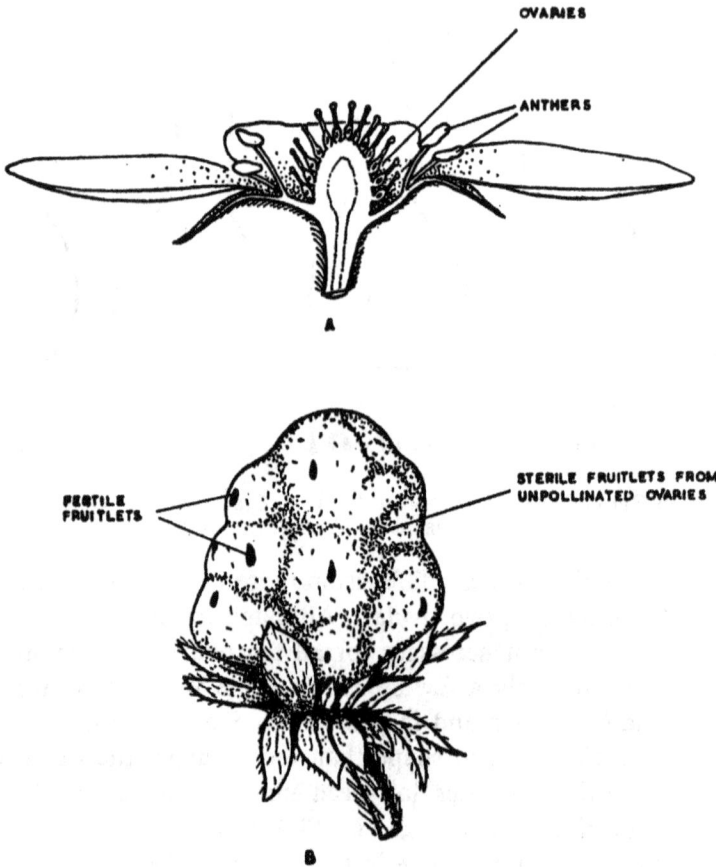

TEXT FIG. 20.
A—Semi-diagrammatic section of a strawberry flower.
B—Malformed strawberry fruit resulting from incomplete pollination.

shed at varying stages of development. Thus, in the apple in a good blossom year, a full crop of fruit may represent only 5 per cent of the total number of flowers on the tree, the remaining 95 per cent having been shed in a series of drops known as the "first" drop, the second or "June" drop, and the "pre-harvest" drop. The first and second drops are basically a manifestation of the intense competition for food materials which exists between the developing fruitlets, but their occurrence is controlled by a specific hormone emanating from the endosperm of the developing seeds. Both these drops occur at particular phases of endosperm development when hormone production is at a minimum, and the fruitlets which are shed have, on the average, a smaller number of seeds—and hence a lower hormone producing capacity—than those which remain attached to the tree (Luckwill, 1948). In addition there are other factors such as spur vigour and position in the cluster which influence the tendency of individual fruitlets to drop (Heinicke, 1917). The relationships between seed number and fruit drop do not hold for the pre-harvest drop of near-ripe fruits later in the season, in which a different hormonal mechanism seems to be operative.

The first drop normally contains a large number of small fruitlets which, owing to defective pollination or to inherent genetical factors, have a low complement of developing seeds. Under conditions exceptionally favourable for full pollination therefore, the first drop tends to be relatively small; subsequent competition between fruitlets, on the other hand, is correspondingly greater, and this leads to an increased June drop. Conversely, where pollination is less thorough, the first drop will be large and the June drop relatively less. These two drops therefore, being complementary, represent a delicately balanced mechanism by which fruit-set is adjusted to suit the food producing capacity of the tree. By virtue of its delicate balance, this mechanism is rather easily upset by outside factors and this may lead either to undercropping, or to overcropping with the consequent production of undersized fruits. Of the outside factors affecting yield by far the most serious is frost, which, by damaging the young seeds, may lead to a great increase in drop or even to total loss of the crop. On certain varieties (*e.g.*, Cox's Orange Pippin) post-blossom applications of lime-sulphur may appreciably increase the first drop without any compensatory decrease of the June drop, thus resulting in decreased yield (Luckwill, 1957b). The mechanism by which lime-sulphur increases fruit drop is not fully understood, though it appears to operate by temporarily reducing the photosynthetic activity of the leaves (Hoffman, 1933), thereby intensifying still further the competition between the developing fruitlets for food materials. Conversely, treatments which increase leaf area at the time of fruit-set (*e.g.*, pre-blossom applications of nitrogen in the form of urea), by increasing carbohydrate production at this critical time may, under certain conditions, reduce the early drop of fruit and lead to increased cropping (Boynton and Fisher, 1949).

Synthetic plant hormones, particularly naphthaleneacetic acid, are now widely used by fruit growers for reducing pre-harvest drop, but they cannot be used for increasing set in the early stages of fruit development. In fact, when applied within three weeks of full bloom they have the effect of inducing seed abortion and, as a consequence, increasing the extent of the early drops of fruit (Luckwill, 1953). Naphthalene acetamide, the most successful hormone thinning agent yet found, is now in general use on apples in the eastern United States, though in this country it has given consistent results only on the variety Crawley Beauty (Luckwill, 1957a).

The above considerations apply primarily to apples. Similar waves of fruit drop are shown by pears and stone fruits and, to some extent by nuts, and although these have not been so intensively studied, it is probable that similar hormonal mechanisms, correlated with the internal development of the seed, operate in these fruits also.

Running-off in Black Currants

Amongst soft fruits the black currant is notable in showing two periods of fruit drop. The early drop, known to growers as "running-off", reaches its peak in the second week after full bloom and continues until the fourth or fifth week. During this period 25 to 50 per cent or more of the developing berries may drop, the extent of the loss depending on such factors as variety, soil, climate and management. In addition to this early period of fruit drop, black currants also show a well-marked harvest drop, which, whilst not normally troublesome, may cause additional losses if picking is for any reason delayed.

Running-off was formerely attributed to inadequate pollination (Wellington et al., 1921) but more recent work (Teaotia and Luckwill, 1956) indicates that this is not the primary cause, and that running-off in currants is of the same general nature as the June drop in apples. As in the apple fruit abscission in the black currant fruit is under the control of a hormone generated in the endosperm of the developing seed (Wright, 1956). The fruits which drop are those with low seed contents which therefore produce less of this abscission preventing hormone than do fruits with larger numbers of seeds. The lower seed numbers of the dropped berries normally result, not from incomplete pollination, but from the abortion of a larger proportion of the fertilized ovules at various stages of development. The cause of this seed abortion is not at present known, but probably hormonal, genetical and nutritional factors are involved. Recent work in Germany (Klämbt, 1958) has shown that seed abortion in black currant and red currant is much reduced when varieties are cross-pollinated and, though self-pollination is normal, this work suggests that the interplanting of different varieties may be helpful in reducing running-off. However, there is as yet no evidence from field experiments that this is so, or that such a reduction in berry drop, if it could be achieved, would result in any increase in crop weight. Attempts to control running-off and harvest drop in black currants with hormone sprays have not led to results of practical significance.

Failure of Fruit-set

The complex processes involved in fruit-setting, as outlined above, are subject to many irregularities, though basically there are only three reasons why a flower may fail to set fruit.

Lack of pollination. This may be due to an absence of pollen (*e.g.*, as caused by male sterility), destruction of the pollen by some external agency (*e.g.*, rain), or to some breakdown in the mechanism by which pollen is transferred to the stigma (*e.g.*, lack of pollinating insects).

Failure of pollen to effect fertilization. This may mean that the pollen grains and/or egg cells are sterile (*e.g.*, owing to irregularities in their chromosome numbers, as in triploids), or that the pollen tubes are incapable of growing down the style (incompatibility).

Seed abortion. Abortion of the young seeds may result from chromosome irregularities, from nutritional causes, or from the effect of external agencies such as frost.

In the paragraphs which follow the numerous factors, both internal and external, which affect fruit-setting are discussed in more detail.

Internal Factors Affecting Pollination and Fruit-setting

COMPATIBILITY

If a sufficient proportion of flowers set fruit with pollen of the same variety for a good crop to result, then that variety is termed self-compatible—or more usually—self-fertile. Varieties that do not set an economic crop with their own pollen are termed partially self-compatible, or, if they fail to set fruit at all, self-incompatible.

Most apple and pear varieities, and all sweet cherries are either self-incompatible or only partially self-compatible, and adequate cross-pollination is essential for satisfactory cropping. With the exception of certain male-sterile varieties, peaches, apricots, quinces, some sour cherries, many commercially important varieties of plums and all soft fruits are considered sufficiently self-compatible to produce good crops without the need for cross-pollination. These different classes are not sharply defined; they are merely convenient groupings based on practical experience, and in fact all degrees of self-compatibility and incompatibility are encountered. Even nominally self-compatible varieites of top fruits may show an increased set of fruit when cross-pollinated and there are indications that the same may be true of soft fruits. Furthermore, the degree of self-incompatibility shown by a particular variety, although determined primarily by the genetic make-up, may be modified within rather wide limits by environmental factors. Thus in some years an apple variety may set a good crop with its own pollen, whereas in other years it will fail to do so (Williams, 1955). In the same way a variety may give consistently good crops without cross-pollination in one locality and yet fail to do so in another.

One factor influencing the degree of self-compatibility of a variety is temperature and its effect on the rate of pollen tube growth, for, under optimal conditions a proportion of nominally incompatible pollen tubes may grow fast enough to reach the ovary before the death of the style. Hence, in general it is found that the self-compatibility of varieites of apple, pear and certain plums tends to be greater in warmer climates or when they are grown under glass. The degree of self-compatibility of a variety also depends on internal physiological factors. Thus, it has recently been shown (Williams, unpublished) that in varieties of apple with marked biennial cropping habits, the self-compatibility is normally greater in the "on" year than in the "off" year.

MALE-STERILITY

The term "sterility" refers to the failure to produce viable pollen grains or egg cells, and must be carefully distinguished from "incompatibility". Sterility is sometimes due to the failure of either the male or female organs of the flower to develop properly, giving rise to male-sterile or female-sterile plants respectively. Female-sterile plants occasionally arise as a result of bud mutation and many ornamental flowering trees (*e.g.*, cherries) show this character. However, since they are unable to set fruit, such plants can obviously never acquire varietal status as fruit trees. Male-sterile varieties, on the other hand, are occasionally encountered amongst cultivated fruits, well-known examples being the pear varieties Marguerite Marillat and Bristol Cross, the plum Golden Esperen, the peach J. H. Hale and the strawberry Tardive de Leopold. Another strawberry variety, Oberschleisien, often shows partial male-sterility. Heslop-Harrison (1957) has found that in certain plants sex-expression is influenced by such factors as light, temperature and mineral nutrition, high levels of nitrogen in particular being effective in promoting femaleness and depressing maleness. High auxin levels in the plant also seem to be associated with suppression of the male organs. This would explain the earlier results of Darrow (1927) who observed that the degree of male sterility shown by certain American strawberry varieties varied from year to year depending on light, temperature and rainfall conditions the previous autumn. It would seem that the tendency to male sterility found in cultivated strawberries may have been inherited from their South American ancestor, *Fragaria chiloensis*, which shows this character to a marked degree. A very unusual type of male-sterility is found in the cider apple varieties Cherry Norman and White Jersey (Williams, 1953). These varieties produce viable pollen grains which, however, are never liberated from the anther owing to a failure of the dehiscence mechanism. All male sterile varieties are, of course, useless as pollinators, though they are usually capable of setting good crops of fruit if cross-pollinated.

STERILITY AND TRIPLOIDS

Some plants have fully developed sexual organs but are rendered sterile by their failure to produce viable pollen grains and egg cells. This type of sterility is particularly characteristic of triploids.

A normal diploid plant has two complete sets of chromosomes in each cell, except in the male and female sex cells, which contain only a single set. The reduction in chromosome number occurs during the final stages of pollen and egg cell development, when the two sets of chromosomes separate, one complete set going into each daughter cell. In triploid plants, with three chromosome sets per cell, such an equal division is not possible and a large proportion of egg cells and pollen grains inevitably receive irregular chromosome complements. This renders them unstable and leads to abortion. It is for this reason that triploid varieties, which are common amongst apples and pears, exhibit a high degree of self-sterility.

The pollen of triploid varieties contains only a small proportion (usually 5 to 20 per cent) of viable pollen grains, and of these only a few may be capable of producing tubes long enough to reach the ovules: they are therefore of little value as pollinators. On the female side however, they are sufficiently fertile to produce good crops when cross-pollinated: indeed, the commercial success of triploid apple varieties such as Bramley's Seedling, Belle de Boskoop and Gravenstein has encouraged breeders in many countries to pay special attention to the production of new triploid varieties. (Plate VIII, Fig. 20).

It may seem strange that a triploid variety that produces perhaps only five per cent of viable pollen grains may nevertheless have a sufficient number of viable ovules to enable it, when cross-pollinated, to set a good crop of fruit. A possible explanation of this is that in each ovule there are four potential embryo-sacs of which only one normally develops. Hence there will be 40 such cells in the whole ovary in apples and pears, and if only five per cent of these develop and are fertilized, this will still give an average of two seeds per fruit. Some ovaries will contain more than this and some less: most of the fruits with two or less seeds will probably drop during development (see p. 131), but there will still remain a sufficient proportion with higher seed numbers to ensure a good crop.

A more puzzling fact, and one not yet fully explained, is that triploid varieties such as Bramley's Seedling may, in some seasons, set a good crop even under conditions which preclude cross-pollination. The seed contents of the fruit is, as might be expected, extremely low, and it would seem that in such seasons there is an increased tendency towards parthenocarpic development of the fruit (see p. 139).

Varieties of apples and pears having four sets of chromosomes per cell are also known. These tetraploids occasionally arise as bud-sports, *e.g.*, Improved Fertility pear (Crane and Thomas, 1939), but more often as chance seedlings in the progeny of triploid varieties (Einset, 1948). In contrast with triploids, these tetraploid forms are highly self-fertile, because at germ-cell formation the four sets of chromosomes divide equally, two complete sets going into each daughter cell to form diploid pollen grains and egg cells. Although the fruits of tetraploids are large, their quality is usually poor and one of their chief uses would seem to be as parents from which, by crossing with diploid varieties, new triploid forms can be produced (Granhall and Oldèn, 1951). Many of the giant bud-sports of apples which have arisen as spontaneous bud mutations are actually chimaeras, having a central core of tetraploid tissue surrounded by one or more cell layers of normal diploid tissue.

INCOMPATIBILITY

Self-incompatibility differs from self-sterility in that viable ovules and pollen grains are produced but, owing to genetic factors, the pollen tubes are incapable of growing down the styles of the variety that produced them. The same pollen however, is capable of growing down the styles of certain other varieties and effecting fertilization. When the pollen of one variety fails to grow in the styles of another variety the term cross-incompatibility is used.

An example will make this clearer:—the cherry variety Early Rivers, like all sweet cherries, is self-incompatibible. If it is self-pollinated the pollen germinates on the stigma and the pollen tubes start to grow down the style; but, after they have penetrated a short way, their growth is arrested so that fertilization does not take place. If Early Rivers pollen is placed on the style of the variety Waterloo, however, it grows normally and brings about fertilization, and in the same way the pollen of Waterloo can fertilize Early Rivers. Cross-incompatibility is also common amongst the sweet cherries. Early Rivers, for example, is cross-incompatible—and therefore useless as a pollinator—with Bedford Prolific, Roundel and at least ten other varieties. Twelve incompatibility groups are recognised in sweet cherries, all the varieties in any one group being cross-incompatible with one another, but cross-compatible with varieties in any of the other eleven groups*.

Incompatibility is determined by the action of genes, in the same way as are the morphological characters of the plant (Crane and Lawrence, 1938). The genes which control incompatibility are designated by the letter S and form a multiple series of alleles S_1, S_2, S_3, etc., any two of which may be carried by a diploid plant. The essential feature of the genetic control of incompatibility is that "like repels like". In other words, pollen cannot function in a style which has the same incompatibility alleles as the pollen itself (Text Fig. 21).

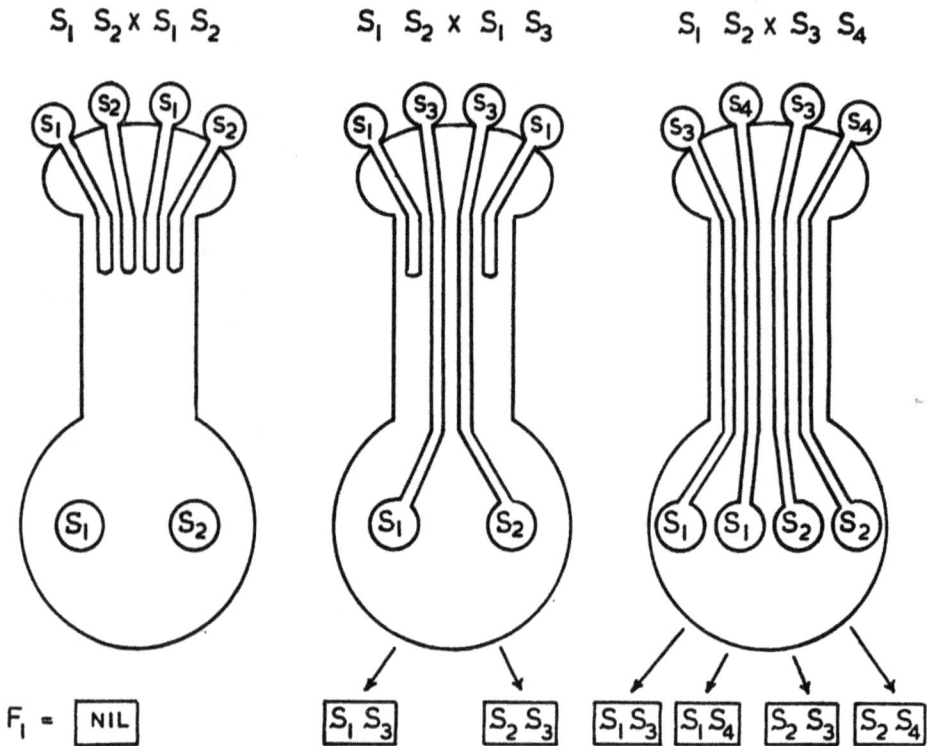

TEXT FIG. 21. DIAGRAM OF POLLEN TUBE GROWTH IN COMPATIBLE AND INCOMPATIBLE POLLINATIONS (AFTER CRANE AND LAWRENCE, 1938).

* These groups will be listed in Part II.

In plums, which are hexaploid, the situation is much more complex than in sweet cherries, since each variety may carry six incompatibility alleles instead of two, and in addition self-compatibility alleles may be present as well. As a result we find amongst the cultivated plums all degrees of self-and cross-incompatibility, ranging from fully compatible to completely incompatible. There are also some which show one-way incompatibility; that is, A pollinated with B sets a full crop, while B pollinated with A fails.

Self-incompatibility also occurs in apples and pears, but is generally less complete than in sweet cherries. In most varieties a few seeds are usually produced from self-pollination, but all varieties set better if they are pollinated with another variety. Cross-incompatibility is not known to occur in apples, but in pears a few cross-incompatible combinations have been recorded (*e.g.*, Williams × Laxton's Superb).

PARTHENOCARPY

Fruits normally depend for their growth on hormonal factors produced in the seeds, but the degree of dependence varies with different fruits. In the strawberry the presence of seeds is required throughout the full growing period of the fruit and their removal at any stage of development leads to an immediate cessation of growth (Nitsch, 1952). In the apple, however, seeds are necessary only up to the end of the "June" drop, their subsequent removal being without effect on the growth rate or ripening of the fruit (Abbott, 1959). Other fruits have carried this tendency still further and have developed the ability to grow without the need for fertilization and seed development. Such fruits, of which the banana is a classical example, are known as *parthenocarpic*, and are characterized by the abnormally high auxin level in their ovaries (Gustafson, 1939).

Amongst fruits grown in this country parthenocarpic tendencies are found in some varieties of apple and pear which, under certain conditions, are capable of setting a crop of seedless fruits. Pear varieties will sometimes set parthenocarpically after the flowers have been exposed to frost, though in Conference and Cheltenham Cross the fruits produced under these conditions are sausage-shaped rather than pyriform. Parthenocarpic fruit-setting after frost damage has also been recorded in apples (Latimer and Rawlings, 1937). The effect of environment on parthenocarpy is well illustrated by the Williams pear which, in this country, will set a crop only after cross-pollination, but which normally develops parthenocarpically in the warmer climate of California (Griggs and Iwakiri, 1954).

Certain apple varieties (*e.g.*, Howgate Wonder) commonly produce crops in which a large proportion of the fruits are apparently seedless. This usually results from the abortion of the seeds at an early stage of development, a phenomenon known as *false parthenocarpy*. Fruits of this type are commonly produced in certain seasons following the self-pollination of triploid varieties.

In some plants such as the tomato parthenocarpic development of the fruit can be induced by auxin sprays, but attempts to do this in the tree fruits have only occasionally proved successful. More recent experiments with apple and pear have shown that in these fruits parthenocarpy may sometimes be induced by sprays of gibberellic acid. The development of a practical method of inducing parthenocarpy in tree fruits would be of great value as a means of ensuring a full crop following frost damage. In strawberries, malformation of the fruit resulting from incomplete pollination can be prevented by spraying with the auxin 2-naphthoxyacetic acid (Swarbrick, 1944).

ANTHER DEHISCENCE

The period of time over which the anthers of a single flower dehisce varies considerably among the different types of fruit. In the currants, which have only four or five anthers per flower, dehiscence is almost simultaneous. Rosaceous flowers have many stamens arranged in whorls which dehisce successively over a period of one to nine days (Table I) though pollen is normally liberated in greater abundance during the first half of this period. If not collected by insects the pollen will remain available and in a viable condition for several days after its release.

TABLE I

DURATION OF ANTHER DEHISCENCE IN INDIVIDUAL FLOWERS

(From data of Percival, 1955)

Fruit	Duration of anther dehiscence (days)
Black currant	Simultaneous dehiscence
Sour cherry	1—2
Strawberry	1—3
Peach	1—5
Apple	1—5
Loganberry	1—6
Pear	2—7
Raspberry	2—9

With the possible exception of the pear, pollen release apparently takes place during the day time only, in some plants (*e.g.*, strawberry) occuring predominantly in the morning and in others (*e.g.*, apple) later in the day (Percival, 1955).

POLLEN QUALITY

Not all the pollen grains shed from the anther are capable of germination, and fruit varieties can be broadly divided into two groups according to the viability of their pollen. Good pollen should show a germination of 70 per cent or higher when cultured in the appropriate concentration of sucrose, but many varieties normally produce pollen with a germination rate of less than 30 per cent (see under Sterility of Triploids). The latter are bad pollinators.

Pollen viability is influenced by many different factors and may vary widely between different flower clusters, between different flowers on the same cluster, and even between different stamens in the same flower. As the pollen becomes mixed during distribution however, such differences are of minor practical significance. Of more importance are seasonal variations in pollen quality, in particular the reduced viability which sometimes occurs following periods of cold, wet weather (Williams, 1953). It has been found that the reduction division stage of pollen grain formation, which occurs at "bud-burst" to "mouse-ear" stage in apples and at "green-cluster" in pears, is particularly temperature-sensitive, and cold weather during this period may adversely affect the quality of the pollen. Pollen quality is also influenced by nutritional factors, particularly the nitrogen status of the parent tree (see p. 146).

DICHOGAMY

Dichogamy, where the male and female organs of the flower mature at different times, is one of nature's devices to encourage cross-pollination. It occurs to a varying degree in both top and soft fruits, which are either pro-tandrous, the anthers maturing first, or protogynous, the stigmas maturing first.

Apples and pears, with some exceptions, are protandrous, and for this reason it has been suggested that for the most effective cross-pollination, a pollinator variety should be chosen that will flower a little after the main variety. However, cross-pollination is a reciprocal process, and with pollinator ratios of 1 : 3— or even 1 : 2—such as are now used in many commercial orchards, this is hardly a sound recommendation if economic crops are to be expected from the pollinator variety, unless the latter possesses a high degree of self-compatibility (*e.g.*, the apple James Grieve).

Some apples (*e.g.*, Allington Pippin, Woolbrook Pippin) and pears (*e.g.*, Conference) show an extreme degree of protandry, the anthers starting to shed their pollen even before the bud opens. Even so, it is unlikely that the degree of dichogamy encountered in top and soft fruits is of any great practical significance, since its effects as a barrier to self-pollination are largely offset by the fact that all the flowers on a tree do not open simultaneously and by the relatively long duration of anther dehiscence.

Protogyny is less common amongst cultivated fruits than protandry. It occurs in certain strawberry varieties (*e.g.*, Deutsch Evern) and in some apples and pears, an extreme example being the perry pear variety Gin in which anther dehiscence may not commence until six days after full bloom.

Dichogamy assumes greatest importance in nuts, in which male and female flowers are borne separately on the same plant. Amongst walnuts there are protogynous and protandrous varieties as well as some in which anthers and stigmas mature at the same time. Many varieties require cross-pollination because, although inherently self-fertile, the degree of dichogamy is too great for satisfactory self-pollination to occur. In the hazel, and probably in other nuts, cold weather retards the development of the male catkins more than that of the pistillate flowers, so that the degree of dichogamy may vary considerably from year to year.

STIGMA RECEPTIVITY AND POLLEN GERMINATION

The stigma is receptive as soon as it becomes covered with the sugary exudate in which the pollen grains can germinate. Stigmas in this condition glisten and remain receptive until they dry and turn brown, the duration of the receptive period depending on variety and climatic factors. In apple and pear not all stigmas in a single flower become receptive at the same time, so that the duration of receptivity is prolonged. In the black currant the stigma become receptive before the petals unfold and remains in this condition for about five days.

Some interesting experiments by Coutaud (1948) with apple pollen suggest that sitgmas secrete certain hormonal substances which reduce the germination of pollen grains of the same variety, but which are comparatively ineffective against pollen of non-related varieties. Thus, it was found that the germination of pollen of Baumann's Reinette on an agar medium was reduced from 89 per cent to 59 per cent by the presence of Baumann's Reinette stigmas, but that it was not affected by the presence of stigmas of Reine des Reinettes. This work, if it were confirmed by more critical observations might throw further light on the problem of self-incompatibility.

In contrast to the inhibiting effect of the stigmas the pollen grains themselves appear to excrete substances (possibly auxins) which stimulate the germination of other pollen grains in the immediate vicinity. This is suggested by the work of Visset (1951) who showed that, when apple pollen is grown in hanging drop cultures of 10 per cent sucrose, the percentage germination varies directly as the number of grains per drop. Recent Russian claims that in many plants improved fruit-setting results from the use of pollen mixtures, as opposed to single-variety pollen, may be related to this phenomenon. Zaec and Sedov (1957) for instance, maintain that in five out of seven apple varieties which they tested the addition of pear pollen to their own pollen improved fruit-set, and that the self-sterile pear Buerré Zimnaya could be made partially self-fertile by the admixture of apple pollen.

In the pear it has been found that the number and rate of growth of the pollen tubes in the style is greatest in the middle of the period over which the stigma is receptive (Modlibowska, 1945), but the chief factor influencing both germination and tube growth is undoubtedly temperature. It is interesting to note that the optimum temperature for the growth of apple pollen lies between 72° and 80°F., which is well above the average temperature experienced at blossom time in this country.

BLOSSOM MORPHOLOGY

In most varieties of apple and pear the stigmas are held slightly above the level of the dehiscing stamens, an arrangement which favours cross-pollination by insect visitors. Varieties such as Cox's Orange Pippin, where the stigmas are well above the level of the anthers, are particularly dependent on insects for pollination and are therefore likely to set poor crops in seasons or situations unfavourable for insect activity. In those varieties where the styles are shorter than the stamens (*e.g.*, Early Worcester, Merton Worcester, Rival), self-pollination even in the absence of insect visitors is facilitated and the chances of cross-pollination are reduced. In Woolbrook Pippin, where short styles are combined with a very early dehiscence of the anthers, self-pollination becomes almost inevitable. Whilst this may have advantages in poor pollination years when insect visitors are scarce, it is not generally beneficial in fruits such as apple and pear where cross-pollination is usually necessary for maximum cropping. Where the variety is highly self-incompatible it may even lead to crop failure, the stigmas becoming so thickly coated with self pollen that foreign compatible pollen is likely to have a reduced chance of germination.

In crops showing a high degree of self-fertility and therefore not dependent on cross-pollination (*e.g.*, black currant) the reverse situation applies, and it is usually considered advantageous to have the stigma level with or slightly below the anthers. In most black currant varieties there is a tendency for a progressive lengthening of the style in successive flowers of the raceme, so that the apical flowers may be more dependent on insect visitors than the basal ones (Wellington et al, 1921).

The relative attractiveness of different flowers to insect visitors is to some extent determined by flower structure, and may vary even between varieties. It has been observed, for instance, that hive bees are more attracted to the flowers of Worcester Pearmain, which has spreading stamens, than to those of Bramley's Seedling in which the close, stout filaments of the upright stamens make the nectaries more difficult of access (Preston, 1948).

BLOSSOM QUALITY AND SPUR VIGOUR

The production of non-viable egg cells by triploids has already been mentioned (page 136), but reduced fertility on the maternal side is also known to occur among diploid varieties. Thus, in the apple Delicious, Hartman and Howlett (1954) found irregular development of the embryo sac existed to such an extent that fruit-set was unfavourably affected. Flowers collected from different positions in the blossom cluster were found to contain varying proportions of ovules with fully differentiated embryo sacs. It is possible that this may help to explain the influence of position in the cluster on the ability of a flower to set fruit; a relationship that has been shown to exist both in certain apples (Howlett, 1931) and pears (Srivastava, 1938).

In the pome fruits, the vigour of the spur, as measured by the size of the bourse and the number of flowers and leaves, can be a factor influencing fruit-set. On weak spurs only those fruits will set which have a high seed content, whereas on vigorous spurs a smaller number of seeds will suffice to prevent abscission (MacDaniels and Heinicke, 1929).

Effect of Pollen Parent on Fruit Characters

Since the growth of the fruit is influenced by hormones emanating from the seeds, the characters of which depend partly on the male parent, it is theoretically possible for the pollen source to influence the growth and form of the fruit indirectly. This phenomenon is known as metaxenia, and it has long been a source of controversy as to whether such effects actually occur amongst the tree fruits grown in this country. The detailed statistical analyses of Nebel (1930) and later of Tydeman (1952) show that the pollen parent can undoubtedly influence size and shape in apples, but that the effects are so small as to be of no practical significance. No influence of pollen source on time of maturity or on fruit quality was detected.

External Factors Affecting Pollination and Fruit-setting

Climatic Conditions

The dependence of pollination and fruit-setting on the weather prevailing at blossom time is universally recognised, as also is the fact that early flowering fruits such as peach and cherry are more likely to be damaged by cold winds and frost than are those which flower later; but no fruits in this country flower sufficiently late to escape entirely these climatic hazards.

In most top and soft fruits frost damage, resulting from ice formation within the tissues, is liable to occur at temperatures below 28°F., but the walnut —the least frost resistant of our fruits—may be damaged at temperatures of 30-31°F. When the temperature of an apple blossom drops to about 28°F., a layer of ice may form below the surface of the receptacle, separating the skin from the cortex (Rogers, 1953). This damage heals, though it may be responsible for later russetting of the fruit. At temperatures of 27°F. or lower the styles and ovules are usually killed, thus preventing pollination. Apples at the early fruitlet stage are even more sensitive to low temperatures than at full bloom, the usual effect of frost at this stage being to kill the developing seeds, thereby inducing fruit drop. Bramley's Seedling is unusual in being particularly sensitive to frost at the "green-cluster" stage, damage being followed by abscission of flower buds or even of the whole cluster. Following damage to the flowers by frost a few varieties of pear (e.g., Durondeau, Conference) are able to set a crop of parthenocarpic fruit, whilst others (e.g., Beurre Hardy, Laxton's Superb) may carry fruits produced from secondary blossom. Mature pollen is unharmed by frost, although after exposure to low temperatures it may take longer to germinate than pollen from undamaged flowers (Field, 1942).

Even when freezing point is not reached, persistent low temperatures at blossom time can have a very drastic effect on fruit-setting. Cold weather prolongs the blossoming season and severely restricts the activity of pollinating insects. Under such conditions a greater overlap in the blossoming period of two varieties is required for adequate cross-pollination, than under good weather conditions. Probably of equal importance is the retarding of pollen germination and tube growth. Apple pollen fails to germinate below 40°F. and pollen tubes make slow growth in the style unless the temperature exceeds 51°F. On the other hand, low temperatures seldom limit anther dehiscence which, in the apple, will occur even at 41°F. (Percival, 1955). Strawberries are exceptional in requiring a temperature of 53°F. or over for anther dehiscence to take place.

Rain may have a deleterious effect on pollination; not only may pollen be washed off stigmas and out of open pollen sacs, but ripe pollen grains, because of the high osmotic concentration of sugar within them, will readily burst in contact with water. Free water on undehisced anthers, however, will prevent the liberation of the pollen, thus protecting it from damage (Percival, 1955).

Wind acts as a pollinating agent of nuts and vines, and possibly assists some soft fruits to be partially self-pollinated. Also, by increasing the concentration of nectar through evaporation it can increase the attractiveness of the flower to insects. However, wind is usually regarded as undesirable during the blossom period, for it dries stigmas, hastens style death and petal fall, and limits insect activity. Windbreaks, by reducing wind velocity, may partially mitigate these effects and can lead to an increased set of fruit (van Rhee, 1959).

AGENTS OF POLLINATION

The only fruit plants known to depend on wind pollination are the cob, filbert and hazel nuts, sweet chestnuts, walnuts and mulberries. The stigmas of other fruit plants are not well adapted to catch wind-borne pollen and investigations have shown that even in windy weather the amount of top fruit pollen in the air within an orchard is negligible. Amongst soft fruits, the pollen of currants and gooseberries is viscid and the grains adhere together in masses, thus preventing wind distribution. The extent to which wind can effect pollination of self-fertile fruit varieties by shaking pollen from the anthers on to the stigmas of the same flower has never received much attention, but there is some evidence that light wind may aid the pollination of certain soft fruits, particularly the strawberry. Overhead watering of strawberries under glass can have a marked beneficial effect on setting and it seems possible that, in the open, rain may also assist strawberry pollination.

The flower structures found in top and soft fruits are clearly adapted to insect pollination and that insects are indeed the chief agents of pollination under field conditions is shown by the many experiments in which fruit-set has been negligible on trees or bushes screened to exclude them, whereas it has been satisfactory on adjacent plants to which insects have had access.

Insects visit the flowers to collect nectar or pollen, or both, and pollen grains trapped in their body hairs may become deposited on the stigma when the insect moves to another flower. A detailed account of the insects which act as pollinating agents for fruit plants will be given in Part 2 of this memorandum.

When fruits are grown under glass, where insects have not free access, hand pollination of the flowers becomes essential. The only exception is the grape, where pollen transfer appears capable of taking place by gravity.

Tree Nutrition

Nitrogen. Nitrogen is one of the most important factors influencing the growth and cropping of fruit trees, but the relationship between the nitrogen level in the tree and its final crop is a complex one. Nitrogen, for example, is a major factor in fruit-bud differentiation, but this subject falls outside the scope of the present review.

The beneficial influence of ground dressings of nitrogen on fruit-set in apples has long been recognised by fruit growers. The effect is probably an indirect one, the principal effect of the nitrogen being to stimulate leaf growth, thereby leading to increased production of carbohydrates which, in turn, assists in the retention of the developing fruitlets through the period of the "June drop". Since the early spring growth of apples, and probably other fruit trees, proceeds at the expense of nutrients accummulated the previous year, the effect of spring dressings of nitrogen on the early development of the spur leaves, and hence on fruit-set, is not as marked as when nitrogen has been applied the previous season. Foliar applications of nitrogen in the form of a pre or post-blossom urea spray, however, may improve the set of apples in some seasons though not in others (Boynton and Fisher, 1949), and in general appear to be uncertain in their effect. Palmiter and Hamilton (1954) found that the incorporation of urea helped to counteract the depressing effect of sulphur sprays on fruit-set of McIntosh apples.

Recent experiments at Long Ashton (Williams, unpublished) have shown that the nitrogen status of the pollen source is also an important factor in fruit-set, at least in some varieties of apples. Pollen from nitrogen-deficient trees, although it germinated well "*in vitro*", was found to be markedly inferior in its capacity to bring about fertilization and fruit-set as compared with pollen from high-nitrogen sources. These results indicate that where the pollinator trees are deficient in nitrogen it is to be expected that the initial set of fruit on the receptor trees will be somewhat reduced, even under favourable pollination conditions, and that the "first" drop will be correspondingly heavy. Pollen from high-nitrogen trees on the other hand, will be effective in bringing about a high initial set and a low "first" drop.

Whereas the extent of the "first" drop, under conditions of full pollination, appears to be influenced by the nitrogen status of the pollen parent, the extent of the "June" drop depends, among other factors, on the nitrogen level in the female parent. In addition, the "June" drop tends to show an inverse correlation with the "first" drop. These relationships between nitrogen and fruit-set may be summarized in tabular form.

TABLE II

		Female parent	
		High N	Low N
Male parent	High N	High initial set Low first drop Moderate June drop	High initial set Low first drop Heavy June drop
	Low N	Low initial set Heavy first drop Low June drop	Low initial set Heavy first drop Moderate June drop

Boron. The pollen grains of most fruit plants will germinate and grow rapidly on an artificial medium containing the appropriate concentration of sucrose. Pear pollen is unusual in that it often fails to grow on such a medium in the absence of an external supply of boron (Schmucker, 1934). The amounts required are minute and usually the addition of 0.1 parts per million of boric acid is sufficient to ensure optimum germination. There is some evidence that the boron requirement of pollen grains from the first flowers to open on a tree is greater than that of the later flowers. Attempts to improve the setting of pears by the application of boric acid sprays have given very inconsistent results and, although boron is undoubtedly an essential factor in pollen germination, there is no evidence that it is ever a limiting factor in fruit set under field conditions in this country.

Magnesium. Although leaf symptoms of magnesium deficiency occur widely in apple orchards and can be readily prevented by magnesium sulphate sprays, there is little published evidence of any crop response from such measures.

However, recent work at East Malling with magnesium-deficient trees of the susceptible scion/rootstock combination 'Edward VII' on 'M.VII', has shown that, as well as preventing leaf scorch, sprays of 2 per cent Epsom salt have also increased the yield by improving the final set (Greenham and White, 1959). The prevention of magnesium deficiency scorch on the small primary leaves of the blossom clusters just after the petal fall stage, appears to be a possible mechanism of this effect.

PRUNING

Pruning by renewal or regulated methods aims at securing blossoms on wood of different ages, normally on one, two, and three-year laterals, with a few blossoms on older spurs. The blossoms on older spurs open first, and are followed by those on three, two and one-year laterals, in that order. Trees that are spur pruned reach their peak of blossoming, as shown by the number of flowers open on any one day, within a few days of the onset of blossoming. Lighter pruned trees have more blossoms open during the latter part of the blossoming period than do spur pruned trees, and this provides a valuable safeguard against poor setting conditions during any particular portion of the period. Conversely, since a spur pruned tree may have about 90 per cent of its blossoms on older spurs, it is possible for these trees to set heavier crops than lighter pruned trees *in some years* when weather conditions happen to be suitable for pollination during only the first few days of the blossom period.

Very heavy winter pruning or dehorning of apples and pears may greatly increase fruit-set, because the large carbohydrate reserves stored in the trunk and roots are then shared amongst a smaller number of fruitlets the following spring. The effect of dehorning on fruit-set may persist for several years Fruit-set may also be increased by branch or trunk ringing which, by breaking the continuity of the phloem, leads to an increased accumulation of carbo-hydrates above the ring.

PESTS AND DISEASES

Almost any fungus or insect pest of fruit will have an adverse effect on cropping, and mention will be made here only of those which, by their attacks on flowers or fruitlets, may directly reduce fruit-set.

The Apple Sucker (*Psylla mali* Schr.) is common in neglected orchards and gardens not subject to an adequate spray programme, and may easily pass undetected. The inconspicuous immature stages of this pest feed throughout the blossom period in the regions towards the base of the flower clusters, and bad infestations may result in almost total failure of fruit-set. By the time this is realised the obvious signs of the pest have disappeared.

Another pest now more frequent in the garden than in the orchard since the advent of DDT is the Apple Blossom Weevil (*Anthonomus pomorum* (L.) Curt.), the larvae of which exert a very direct effect on fruit-set by eating the sexual organs within the enclosed petals, which never expand beyond the balloon stage. A related weevil causes similar damage to strawberry flowers. Other pests whose presence is rendered obvious by the early shedding of fruitlets include the Apple Sawfly (*Hoplocampa testudinea* Klug.), whose larvae feed on the young seeds, the Plum Sawfly, and the Pear Midge (*Contarinia pyrivora* Riley).

Apple Mildew (*Podosphaera leucotricha* (Ell. & Ev.) Salm.), which in recent years has increased in importance in this country, overwinters on the outside of the shoots and also within the buds. Infected fruit-buds in spring give rise to small and malformed blossom clusters which usually fail to set fruit. A disease which can attack the flowers of practically all top fruits, but which is most commonly met with on plum, is Blossom Wilt (*Sclerotinia laxa* Aderh. and Ruhl.). This is most prevalent in wet seasons and causes complete death of the flowers. Grey Mould (*Botrytis cinerea* Pers.) and Mildew (*Sphaerotheca humuli* (DC.) Burr.) are two diseases often responsible for the destruction of a proportion of strawberry flowers.

References

ABBOTT, D. L. (1959). The effects of seed removal on the growth of apple fruitlets. *Ann. Rep Long Ashton Res. Sta. for* 1958, 52-56.

BISHOP, C. J. (1953). The inheritance of tree and fruit characters in natural polyploid apple seedlings. *Proc. Amer. Soc. hort. Sci.*, **62**, 327-333.

BOYNTON, D. and FISHER, E. G. (1949). Results of use of urea sprayed on foliage. *Proc. N.Y. hort. Soc.*, **94**, 101-103.

COUTAUD, J. (1948). Influence des stigmates sur la qualité germinative des pollens de différentes variétiés de pommier. *Comptes Rend. Acad. Sci.*, **226**, 2090-2091.

CRANE, M. B. and LAWRENCE, W. J. C. (1938). *The genetics of garden plants.* Macmillan, London.

CRANE, M. B. and THOMAS, P. T. (1939). Genetical studies in pears I. The origin and behaviour of a new giant form. *J. Genetics*, **37**, 287-299.

DARROW, G. M. (1927). Sterility and fertility in the strawberry. *J. agric. Res.*, **34**, 393-411.

EINSET, J. (1948). The occurrence of spontaneous triploids and tetraploids in apples. *Proc. Amer. Soc. hort. Sci.*, **51**, 61-63.

FIELD, C. P. (1942). Low temperature injury to fruit blossom. II. A comparison of the relative susceptibility and effect of environmental factors on three commercial apple varieties. *Rep. E. Malling Res. Sta. for* 1941, 29-35.

GRANHALL, I. and OLDÉN, E. J. (1951). De tetraploida äpplenas utnryttjande i växtförädlingsarbetet. *Sver. Pomol. Fören Arsskr.*, **52**, 47-65.

GREENHAM, D. W. P. and WHITE, G. C. (1959). The control of magnesium deficiency in dwarf pyramid apples. *J. hort. Sci.*, **34**, 238-247.

GRIGGS, W. H. and IWAKIRI, B. T. (1954). Pollination and parthenocarpy in the production of Bartlett pears in California. *Hilgardia*, **22**, 643-678.

GUSTAFSON, F. G. (1939). The cause of natural parthenocarpy. *Amer. J. Bot.*, **26**, 135-138.

HARTMAN, F. O. and HOWLETT, F. S. (1954). Fruit setting of the Delicious apple. *Bull. Ohio agric. Expt. Sta.*, 745.

HEINICKE, A. J. (1917). Factors influencing the abscission of flowers and partially developed fruits of the apple (*Pyrus malus* L.). *Bull. Cornell agr. Exp. Sta.*, 393.

HEINICKE, A. J. (1927). The set of fruit as influenced by pruning at different periods preceeding the bloom. *Proc. Amer. Soc. hort. Sci.*, **23**, 46-48.

HESLOP-HARRISON, J. (1957). The experimental modification of sex-expression in flowering plants. *Biol. Revs.*, **32**, 38-90.

HOFFMAN, M. B. (1933). Carbon dioxide assimilation by apple leaves as affected by lime-sulphur sprays. II. Field experiments. *Proc. Amer. Soc. hort. Sci.*, **30**, 169-175.

HOWLETT, F. S. (1931). Factors affecting fruit setting. I, Stayman Wine-sap. *Bull. Ohio agr. Exp. Sta.*, 483.

KLAMBT, H. D. (1958). Unteruschungen über die Befruchtungs-verhältnisse bei Schwarzen und Roten Johannisbeeren. *Gartenbauwiss.*, **23**, (5), 9-28.

LATIMER, L. P. and RAWLINGS, C. P. (1937). The occurrence of seedless apples as a result of frost. *Proc. Amer. Soc. hort. Sci.*, **35**, 111-115.

LUCKWILL, L. C. (1948). The hormone content of the seed in relation to endosperm development and fruit drop in the apple. *J. hort. Sci.*, **24**, 32-44.

LUCKWILL, L. C. (1949). Fruit development in relation to plant hormones. *Endeavour*, **8**, 188-193.

LUCKWILL, L. C. (1953). Studies on fruit development in relation to plant hormones. II. The effect of naphthalene acetic acid on fruit-set and fruit development in apples. *J. hort. Sci.*, **28**, 25-40.

LUCKWILL, L. C. (1957a). The chemical thinning of apples in Britain. *Agric. Rev.*, **3**, 19-22.

LUCKWILL, L. C. (1957b). Fruit-set on Cox's Orange Pippin in relation to spray programme. *Ann. Rep. Long Ashton Res. Sta. for 1956*, 61-65.

MACDANIELS, L. H. and HEINICKE, A. J. (1929). Pollination and other factors affecting the set of fruit, with special reference to the apple. *Bull. Cornell agr. Exp. Sta.*, 497.

MODLIBOWSKA, I. (1945). Pollen tube growth and embryo-sac development in apples and pears. *J. hort. Sci.*, **21**, 57-89.

MURNEEK, A. E. (1939). Further results on the influence of branch ringing on fruit set and size. *Proc. Amer. Soc. hort. Sci.*, **36**, 398-400.

NITSCH, J. P. (1952). Plant hormones in the development of fruits. *Quart. Rev. Biol.*, **27**, 33-57.

NEBEL, B. R. (1930). Xenia and metaxenia in apples. *Geneva Tech. Bull.* 170.

PALMITER, D. H. and HAMILTON, J. M. (1954). Influence of certain nitrogen and fungicide applications on yield and quality of apples. *Bull. Cornell agr. Exp. Sta.* 497.

PERCIVAL, M. S. (1955). The presentation of pollen in certain angiosperms and its collection by *Apis mellifera*. *New Phytol.*, **54**, 353-368.

PRESTON, A. P. (1949). An observation on apple blossom morphology in relation to visits from honey bees (*Apis mell fera*). *Rep. E. Malling Res. Sta. for 1948*, 64-67.

VAN RHEE, J. A. (1958). The cropping of fruit trees in relation to windbreak protection. *Netherlands J. agric. Sci.*, **6**, 11-17.

ROGERS, W. S. (1953). Some aspects of spring frost damage to fruit and its control. *Proc. 13th Int. Hort. Cong. 1952*, 941-947.

SCHANDER, H. (1955a). Untersuchungen über die Gestalt der Frucht bei Kernobst. *Gartenbauwiss.*, **1**, 313-324.

SCHANDER, H. (1955b). Uber die Veränderlichkeit der Fruchtgestalt bei der Birnesorte "Conference". *Mitt. ObstVersAnst. Jork.*, **10**, 271-277.

SCHANDER, H. (1956). Uber die Ursachen von Gewichtsunterschieden bei Samen von Kernobst (Apfel und Birne). II. Der Einfluss verschiedener Erb-und Umweltfaktoren auf die Beziehungen zwischen Samen und Frucht. *Zeit. für Pflanzenzüchtung*, **36**, 31-80.

SCHMUCKER, T. (1934). The influence of boric acid on plants, especially germinating pollen grains. *Planta.*, **23**, 264-83.

SRIVASTAVA, D. N. (1938). Studies in the non-setting of pears. I. Fruit drop and the effect of ringing, dehorning and branch bending. *J. hort. Sci.*, **16**, 39-62.

SWARBRICK, T. (1944). Progress report on the use of naphthoxyacetic acid to increase fruit-set of the strawberry variety Tardive de Leopold. *Ann. Rep. Long Ashton Res. Sta. for 1943*, 31-32.

TEAOTIA, S. S. and LUCKWILL, L. C. (1956). Fruit drop in blackcurrants: I. Factors affecting "running-off". *Ann. Rep. Long Ashton Res. Sta. for 1955*, 64-74.

THOMPSON, A. H. (1957). Chemical thinning of apples. *Bull. Md agric. Exp. Sta.*, A88.

TUKEY, H. B. (1936). A relation between seed attachment and carpel symmetry and development in *Prunus*. *Science*, **84**, 513-515.

TYDEMAN, H. M. (1952). The influence of the pollen parent on the development of apple fruits. *Rep. E. Malling Res. Sta. for 1951*, 62-66.

VISSER, T. (1951). Bloembiologie en kruisingstechniek bij appel en peer. *Meded. Dir. Tuinb.*, **14**, 707-726.

WELLINGTON, R., HATTON, R. G. and AMOS, J. (1921). The "running-off" of blackcurrants. *J. Pomol.*, **2**, 160-198.

WILLIAMS, R. R. (1953). Pollination requirements of cider apple varieties. Progress Report 1952. *Ann. Rep. Long Ashton Res. Sta. for 1952*, 44-48.

WILLIAMS, R. R. (1955). Pollination requirements of cider apple varieties. II. Progress report, 1954. *Ann. Rep. Long Ashton Res. Sta. for 1954*, 38-46.

WRIGHT, S. T. C. (1956). Studies of fruit development in relation to plant hormones. III. Auxins in relation to fruit morphogenesis and fruit drop in the blackcurrant *Ribes nigrum*. *J. hort. Sci.*, **31**, 196-211.

ZAEC, V. K. and SEDOV, E. N. (1957). The effect of conditions of growth on the selectivity of fertilization in apples (Russian). *Agrobiologija*, **3**, 67-71.

PART II

POLLINATION METHOD

The Insects Concerned in Pollination

Insects, as the agents of pollen transfer, play a vital part in the success of the fruit growing industry yet, in this role, surprisingly little information concerning them is readily available. Some information relating to the kinds and species involved under European conditions has nevertheless been gathered, notably by Hooper (1920), Wellington *et al.* (1921) and Fox Wilson (1929) in England, and by Stapel (1939) in Denmark and Faber (1953) in Western Germany. Our knowledge is far from complete, however, and this section claims no more than to bring together some information about those insects believed at present to be of value.

HONEY BEES

The honey bee (*Apis mellifera L.*), and other races of hive bees, depend for their existence on the pollen and nectar they collect from flowers. Pollen becomes trapped in the branched hairs which cover their bodies, and although the bee cleans its body of pollen which it places in pollen baskets on its rear legs, many scattered pollen grains are left. It is this remaining pollen that effects pollination.

A field force of foraging bees is seldom engaged in a single activity, being normally divided into pollen gatherers, nectar gatherers, and those gathering both pollen and nectar. Honey bees foraging for pollen usually clamber over the stamens, touching the stigma as they do so, but on some apple and plum varieties nectar gatherers often alight on the petals, and by probing between the filaments from the side, extract nectar without coming into contact with either anthers or stigma (Preston, 1949; Brown, 1951). Because the activity of pollen collection necessarily brings the honey bee into contact with the stigma, and as many workers have observed on various top fruits that pollen gatherers work faster than nectar gatherers, the former may be regarded as the more efficient pollinators. The proportion of bees engaged in each activity varies greatly from day to day in relation to the availability of pollen or nectar and to the needs of the colony.

Although the flight range of the hive bee may vary, Percival (1947) found the normal limit for pollen collection to be a quarter of a mile. Similar distances for flight range in apple orchards were found by Brittain *et al.* (1933) and Momers (1948a), though where nectar is scarce hive bees may travel twice this distance or even further (Vansell, 1942). The area within flight range of a single colony or group of colonies is thus approximately 120 acres. The significance of this is shown by the work of Brittain *et al.* (1933), who found that the density of bees on apple trees where hives were placed at one per acre was very much reduced when the orchard in which the hives were placed was surrounded by, or adjacent to, further orchards in which hives of bees had not been placed.

While studying the foraging areas of bees on apples, Momers (1948b) found that where trees were planted closer in one direction than the other, hive bees tended to fly much more along than between the rows, even though the inter-row distance was short. Similar observations have been reported by Singh (1950) and Rymashevskii (1956). This preference for the shortest route between trees probably explains the effect of internal roadways in orchards which, by suddenly increasing the tree to tree spacing, tend to reduce bee activity in that portion of the orchard furthest from the hives (Momers, 1948b; Whiffen, 1948).

Individual honey bees do not forage over the whole area within flight range, but restrict their activities to a small portion of it, to which they repeatedly return. Several workers have shown on herbaceous crops that such foraging areas, while variable in size, may be only a few feet in diameter. From this it would appear that the foraging area of an individual bee is closely related to the area of flowers required to yield a series of pollen or nectar loads; consequently on large trees carrying many thousands of blossoms, a bee need only visit a small fraction of them to satisfy her requirements. Observations on pears support this; Vansell (1942) found that individual honey bees were able to collect a load of pollen or nectar from as few as 84 flowers. It is hardly surprising, therefore, that on apple trees of 20-25 feet in diameter which probably had many thousands of open blossoms, Singh (1950) found that of a sample of 66 hive bees, two-thirds foraged only on one tree or part of a tree, and that most of the rest worked between two trees. The restriction of foraging, usually to a single tree, is confirmed by the observations of MacDaniels (1931) on apples and by Stephen (1958) on pears. Stadhouders and Momers (Momers, 1948b), however, working with young dwarf apple trees (spindle-bushes) found that one individual tree might be visited during the course of a day by many bees, and that while these undoubtedly foraged over a restricted area, this included several trees, some being up to 90 feet apart. Work carried out by the same workers early in the blossom period with fewer colonies (two per acre) indicated even greater overlapping between the foraging areas of different bees.

It would appear that the conditions under which Stadhouders and Momers carried out their experiments provided a high bee/blossom ratio and this was possibly responsible for the extensive overlapping of foraging areas that they found. Nevertheless, the relatively small amount of forage that individual dwarf trees provide, by necessitating the inclusion of several trees within the foraging areas of individual bees, must be conducive to cross-pollination. Hence size of tree may well be one factor determining the effectiveness of honey bees as cross-pollinators. Further, since at the beginning and end of the flowering period the amount of open blossom is relatively less and the insect/blossom ratio is therefore presumably higher, the chances of cross-pollination may be greater at these times than at full bloom.

To account for the transference of pollen between large trees, Singh (1950) suggests that as the foraging areas of certain bees may embrace areas of blossom on two trees, the pollen that has been transferred from one tree to another may then be more widely spread over the individual trees by other bees. However, this does not seem to be the case, for Johansen and Degman (1957) were unable to obtain any redistribution of foreign pollen on a partly hand-pollinated apple tree enclosed in a cage with a hive of bees, and the observation that fruit-set can be highest on the side of the tree adjacent to the pollinator (Brittain *et al.*, 1933; Brown, 1951) indicates that an efficient secondary distribution of pollen cannot be relied upon. Quite apart from the redistribution of pollen, it is also pointed out by Singh that the number of occasions when an individual bee visits more than one tree may be greater than observations indicate. Foraging areas are known to change under the influence of a variable supply of nectar and pollen and disturbance by wind and other insects, and even bees whose normal areas is restricted to one tree may move for a time to others, and so bring about cross-pollination.

However, Butler (Butler and Simpson, 1954) considered that honey bees with restricted foraging areas could not be responsible for cross-pollination in orchards. Since the necessary transfer of pollen does occur, Butler, assuming honey bees to be the pollinating agents responsible for this, postulated the existence of an additional population composed mainly of young "wandering bees" which have not yet found satisfactory foraging areas and which presumably visit those of other bees while searching for one of their own.

Another theory offering a possible explanation of the mechanism of cross-pollination in orchards is the transfer of pollen from one bee to another within the hive. It would appear, however, that such a mechanism can hardly be important even if it exists, for, if it were so, and assuming that honey bees are important pollinating agents, the reduction in yield which accompanies increasing distance from a pollinator would not take place (see Tables VI-VIII), and blocks of pollinators in one part of an orchard would prove effective.

BUMBLE BEES

Several species of bumble bee have been found in orchards and plantations at blossom time. Among the commonest are *Bombus lapidarius* (L.), *B. terrestris* (L.) and *B. lucorum* (L.). A late emerging species, *B. helferanus* Seidl., has been recorded as a frequent visitor to raspberry and loganberry flowers.*

Although no precise studies have been made of the foraging behaviour of bumble bees on fruit plants, Brown (1951) and Menke (1951) both agree that bumble bees fly from tree to tree more readily than do honey bees and are thus potentially better cross-pollinators. Bumble bees are large, hairy, and so make contact with the stigma on most visits whether collecting nectar or pollen;

---

* Information contained in the passages marked with an asterisk is derived chiefly from the papers of Faber and Fox Wilson.

indeed, Brown considered that the flowers of certain varieties, particularly the plum 'President', could only be effectively pollinated by large insects such as bumble bees on account of the considerable distance beyond the stamens to which the stigma extends. Wellington *et al.* (1921) observed that the visits of bumble bees to black currants caused the whole raceme to shake vigorously, and pointed out that this action alone might lead to useful self-pollination. The frequency of blossom visitation varies with the species concerned, but whenever bumble and honey bees have been compared on the same plant, bumble bees have always been found to visit the larger number of flowers per minute.

These characteristics are valuable attributes for, unlike honey bees, most bumble bees seen during the fruit blossom period are queens which have over-wintered, and hence the number of individuals available for pollination is relatively small. Attempts to increase the population of bumble bees, if success-ful, would seem worthwhile. Several practical possibilities exist and these are discussed in some detail by Free and Butler (1959). Recommendations for the conservation of wild pollinating insects in general are also given by Bohart (1952).

OTHER WILD BEES

Osmia rufa (L.), several species of *Andrena* and two of *Halictus* have been recorded as visitors to fruit blossom and may be valuable pollinating agents as they spend most of their foraging time collecting pollen rather than nectar. Studying the behaviour of several species of *Andrena* on fruit trees, Chambers (1946) concluded that where abundant, *A. varians* Rossi., could be of considerable value as a pollinator as it showed greater preference for fruit pollens than the other species observed and its foraging area was not restricted to individual trees. According to Fox Wilson (1933) some species such as *A. haemorrhoa* (Fab.), are commonly met with on both top and soft fruits whereas *A. thoracia* (Fab.), has only been recorded on bush and cane fruits.

Under Danish conditions, Stapel (1939) found *Andrena* species slow working, visiting on average only half as many flowers per minute as honey bees. Pollen adheres more loosely to wild bees and consequently it is more easily brushed off during visits to a flower; it is also carried on a larger area of their bodies (Brittain *et al.*, 1933). Thus it is possible that the disadvantage of their com-parative slowness of working may be partially offset by increased pollinating efficiency.

FLIES

Midges or fungus gnats may on occasion far surpass in numbers all other pollinating insects on apples, pears, plums and cherries. While many are very small and their visits sporadic, Fox Wilson (1929) considered them capable of distributing pollen. They are found from early morning to late at night amongst flowers, and are present in stormy weather when very few other insects are seen.

The two fever flies, *Bibio hortulanus* (L.) and *B. marci* (L.) (the St. Mark's Fly) and the more prevalent *Dilophus febrilis* (L.) may be considered useful pollinating agents of top fruit.*

Hover flies, species of *Eristalis* and *Syrphus*, are constant visitors to the flowers of top and soft fruits. However, a large amount of time is spent hovering above trees and bushes, particularly under warm conditions, and their efficiency as cross-pollinators is further lowered by the fact that they will often revisit the same flower many times. Indeed, there is evidence that certain species at least show almost the same restriction to small foraging areas as honey bees (Minderhoud, 1951). Several other members of the Syrphidae are to be found on fruit blossom, for example *Rhingia rostrata* (L.) may be mentioned as a frequent visitor to the flowers of top fruits. Momers (1948a) observing the habits of *R. rostrata* and *Tubifera tenax* (L.) on apple, found that their bodies seldom make contact with the anthers or stigmas and that compared with honey bees, very little pollen could be found on them. In general, therefore, syrphids, although often present, are probably of minor value as pollinators.

The dung fly *Scatophaga stercorarium* (L.) is of interest as being found particularly on pears.*

Several small flies, members of the Anthomiidae, are present in large numbers on blossoms in orchards. In addition, blowflies such as *Calliphora erythrocephala* (Meigen) are considered useful pollinators,* particularly of pear (Bohart, 1952) and plum (Brown, 1951) to which they seem attracted, though this view is not shared by Vansell (1942) who observed blowflies in a Californian pear orchard. He considered that their habits of walking about the trunk, branches and leaves as well as the blossoms was not conducive to effective pollination.

BEETLES

Many observers regard the various beetles and weevils found in fruit blossoms to be of no consequence in pollination. Massee (1937) however, doubts this. He noted that numerous small flower beetles, representing many species, were to be seen in the flowers of fruit trees and that often they were covered with pollen. He pointed out that they have the valuable characteristic of remaining in the blossoms irrespective of prevailing weather conditions, and that as they fly readily, they presumably fly from one tree to another in the orchard.

NOCTURNAL VISITORS

An economic set of fruit following a pollination period during which few insects of any sort have been observed, has led, from time to time, to the speculation that night flying insects might be responsible. Under conditions in the Altenland district of Germany in 1951, however, Faber (1953) found that there were so few visitors to fruit blossoms at night when it was usually cold, that these could not be regarded as important, particularly as their movements were sluggish due to the low night temperatures; Fox Wilson (1926) also searched for moths on the flowers of fruit plants at night, but found none.

Insect Behaviour

CONSTANCY OF VISITING

Some nectar and pollen gathering insects visit a range of plant species in one trip. Honey bees, however, show a high degree of constancy to one crop at a time, a feature that renders them particularly useful as pollinators. Faber (1953) examined several kinds of wild insects to see if they resembled honey bees in this respect. He identified the pollen taken from species of *Bombus*, *Andrena*, *Osmia*, *Eristalis intricarius* (L.) and *Dilophus febrilis* (L.), and, finding a high percentage of fruit pollen in each case, concluded that there was no reason to doubt the usefulness of these insects as pollinating agents.

INSECT ACTIVITY IN RELATION TO WEATHER

Weather plays a large part in governing the number and kinds of insects visiting fruit blossoms.

Low temperatures, lack of sunshine, wind and rain all reduce the number and kinds of insects visiting fruit blossom. The relative disregard that bumble bees, and to some extent midges, have for inclement weather is clearly shown (Tables III and IV). Low light intensity is the most important factor limiting the activity of bumble bees which can often be found in the orchard or plantation from soon after sunrise until just before sunset. Within the temperature range of honey bee activity, Brittain *et al.* (1933) found that light apparently has a more important influence than slight changes of temperature, a reduction in light intensity causing a reduction in bee activity; this may explain why hive bees seldom work flowers which are in partial shade. The optimum temperature for hive bees and the species of *Halictus* and *Andrena* found on apple in Nova Scotia, was considered by Brittain *et al.* (1933) to be approximately 68°F., but the temperature below which bees will not forage is variable. Rashad (1957) found that in spring hive bees collected pollen at temperatures as low as 46-52°F., but it is known that at low temperatures foraging is confined to those bees whose foraging areas are close to the hive. It is thus probable that until the temperature rises above 60°F. foraging activity remains considerably restricted. The limitations placed by light intensity and temperature on the flight of honey bees mean that on many days which are otherwise conducive to the visitation of fruit blossom, they may only forage in strength during late morning and early afternoon. Wind speeds above 11 m.p.h. were found by Rashad to reduce the activity of pollen gathers, while at 21 m.p.h. the bees remain in the hive. If sufficiently strong, wind also influences the direction in which bees forage. Bumble bees and some species of *Andrena* are much less sensitive to wind than honey bees.

Weather outside the blossom period can also exert a large influence on the force of pollinating insects at blossom time. Excessively wet weather may cause flooding of the overwintering quarters of those wild species which hibernate underground, while mild spells in early spring followed by severe cold can also greatly deplete the population.

TABLE III

THE EFFECT OF WEATHER ON THE VISITS OF INSECTS TO APPLE BLOSSOM*

	April 15, 1921, 2.45-3.15 *p.m.* snow showers with sunny intervals, slight N. wind. 38°F.	*April* 30, 11.15-11.45 *a.m.* bright sun slight N. wind. 64°F.	*May* 4, 11-1.30 *a.m.* overcast with showers, slight S.E. wind. 43.5°F.	*May* 11, 11-11.30 *a.m.* overcast and close, slight E. wind. 59°F.
Hive bees ...	0	17	0	0
Bumble bees ...	1	5	3	14
Wild bees ...	0	21	0	0
Hover flies ...	0	13	0	1
Fever flies ...	0	1	0	2
Midges ...	7	very numerous	10	numerous
Anthomyiids ...	0	8	0	0
Flies ...	0	4	0	0

* from Fox Wilson (1926)

TABLE IV

OBSERVATIONS ON A SINGLE BLACKCURRANT BUSH FOR ONE HOUR ON THREE CONSECUTIVE DAYS*

	May 4, 1920, 9.15-10.15 *a.m.* N.W. breeze, moderately cloudy with sun obscured at intervals	*May* 5, 9.15-10.15 *a.m.* S.W. breeze, very few clouds, sun strong	*May* 6, 10-11 *a.m.* S.W. wind, very cloudy, with rain and no sun
Hive bees	2	23	0
Bumble bees ...	3	3	1
Wild bees	5	12	0
Flies (including Fever flies) ...	26	46	5
Midges	1	3	0
Ants	9	5	3

* from Wellington et al (1921).

THE IMPORTANCE OF NECTAR IN ATTRACTING INSECTS

Quantity of Nectar.—Brown (1951), examining certain aspects of the pollination of plums, found that the quantity of nectar produced varied greatly between varieties, some secreting ten times as much as others. Although the concentration of these nectars was not recorded, Brown noted that those varieties with the greatest quantity of nectar attracted the largest number of bees and that these varietal preferences persisted even when foraging activity changed from nectar to pollen collection. This observation agrees with those of Percival (1947) who concluded that if an attractive source of nectar is available, honey bees will also make it the chief source of pollen. Thus although nectar gatherers may themselves be less efficient pollinators, nectar is important as an indirect means of attracting pollen gatherers.

Nectar Concentration.—When several kinds of plants flower simultaneously, pollinating insects are more attracted to some than others. In general, the attractiveness of flowers to honey and bumble bees closely corresponds to the sugar concentration in the nectar, and this varies over a wide range (Table V). There will therefore be competition for insect visitors among plants in flower at the same time. Insects may be attracted from one fruit crop to another or away from fruit entirely. The concentration, and consequently the attractiveness of nectar in any one flower, varies with environment, particularly humidity, and the exposed nectaries of many fruit plants also make nectar dilution with dew and rain a common occurrence. Subsequent evaporation to a higher concentration may re-attract the attention of insects, however. Although variability due to environment and variety limits the use that can be made of determinations of the average sugar concentration in nectars, Table V nevertheless indicates why honey bees may be often noticeably scarce in plum and pear orchards.

TABLE V

AVERAGE SUGAR CONCENTRATION (PER CENT.) IN NECTAR OF VARIOUS FRUITS

Sweet cherry (1)	...	...	55	Sour cherry (1)	28
Apple (2) ...	...	...	42	Blackberry (2)	28
Raspberry (2)...	...	...	37	Blackcurrant (2)	25
Strawberry (3)	...	...	30	Plum (2)	21
Peach (1) ...	...	...	30	Pear (2)	15

(1) Vansell (1942)   (2) Butler (1945)   (3) Shaw (1954)

Two clear examples of competition have been given by Vansell (1952). He observed bees to visit sweet cherries (nectar 55 per cent sugars) but not sour cherries growing in the same orchard (nectar 20 per cent sugars), and cover crops of mustard (nectar 44-60 per cent sugars) were noted to attract bees away from plum and pear (nectar 12 per cent sugars). Almost certainly a similar explanation can be given for the desertion of currants and gooseberries observed by Fox Wilson (1926), when nearby cherries were in full flower, and for the differential attractiveness reported between varieties of plums (Brown, 1951), pears (Vansell, 1942) and also of apples (Overley *et al.*, 1946). These latter authors found that under their conditions hive bees tended to concentrate on one variety until its nectar was exhausted before turning to another later in the day, and pointed out the disadvantage of choosing a pollinator whose attractiveness to bees was less than that of the main variety. Cooper (private communication) observed that honey bees, working blackcurrant in relatively inclement weather, flew further afield to brassica seed crops as the weather improved later in the day. Bees are often attracted to the flowers of dandelions which secrete nectar at lower temperatures than the apple. Dandelion flowers close under rapidly rising temperature conditions and thus limit competition when it is really warm, but such conditions do not predominate during the fruit blossoming period in this country, and the presence of dandelions in the orchard sward is undesirable.

The Use of Sugar Sprays.—As plum and pear varieties tend in general to have low nectar concentrations, attempts have been made to increase their attractiveness to bees by sugar sprays; the results, however, are conflicting. Roberts (1956) found them successful on plums, increasing the number of bees per tree from 6 to 500, whereas similar measures carried out on pears (Stephen, 1958) not only failed to attract an increased number of bees, but actually reduced the number on the flowers, as some bees collected sugar from leaves and twigs.*

The possibility of attracting bees to particular crops by feeding colonies with syrup scented with the flowers of the desired plants has received considerable attention. Such a technique, were it successful, would be of obvious value. In the carefully conducted experiments carried out by Free (1959), however, the feeding of scented syrup had little or no effect on the amount or proportions of pollen collected from the two crops concerned—red clover and apple. As it is during pollen collection that most pollination is carried out, Free concluded that these results gave no indication that scent-direction is likely to become a useful practice.

THE INSECT TO BLOSSOM RATIO

To some extent the picture which insect pollination presents contains so many limitations that it may seem remarkable that sufficient cross-pollination is ever achieved. In general, however, the average yields of our various fruits are high enough to leave little doubt that sufficient pollination is, in fact, usually effected. It is easier to understand this, perhaps, when it is realized that the magnitude of the task is not as great as it might appear.

---

* Fruit blossom can be made unwittingly less attractive to bees by spraying during the blossom period with certain materials, *e.g.*, lime-sulphur, which may act as temporary repellents. Not all sprays are dangerous to bees, but lead arsenate and BHC should be avoided during the blossom period or large numbers of both hive bees and wild pollinating insects may be killed.

An economic yield from an acre of mature apples is approximately 55,000 fruits, representing roughly 5 per cent of the flowers present in a good blossom year. While it is known that the numbers of flowers visited by insects in a given time differs between the kinds of insects and under varying conditions, a frequency of 5 flowers per minute would be a low estimate for the larger insect visitors. To maintain a continuous frequency at this level for 5 hours a day, while it would require more than one insect, would involve the visitation of 1,500 flowers. Thus the 55,000 flowers to be pollinated could be visited in 5 hours by a population capable of maintaining 37 individuals constantly on the blossom. As an acre of mature apple trees can produce roughly one million flowers this means, assuming cross-pollination to be effected at each visit to a flower, that one insect per 30,000 blossoms could set a crop in a single day. While it is highly probable that cross-pollination never reaches maximum efficiency under natural conditions, these theoretical calculations do at least suggest that fewer pollinating insects are required than is commonly supposed. The fact that good fruit setting has been known to occur in season when few insects have been seen lends support to this view.

Relative Values of Wild Insects and Hive Bees

Evidence that naturally occurring populations of insects can be valuable pollinators of fruit has been provided by several observers. Thus, excellent crops of fruit were harvested in 1917 in Essex and in 1920 in Surrey when hive bees were observed to be completely absent (Grainger, 1929; Fox Wilson, 1929), and Loken (1958) who carried out a detailed study on marked branches of the apple Gravenstein in Norway, observed a satisfactory set in two out of three years when honey bees were completely absent, the chief pollinating insects being bumble and solitary bees. In New England, Sax (1922) stated that good apple crops had been obtained where bumble bees were the only apparent pollinating agent, and Brittain et al. (1933) considered that in parts of Nova Scotia where there was a virtual absence of both bumble and honey bees, the natural population of Halictus and Andrena species were responsible for the successful pollination of apple. Unfortunately few investigations have been carried on for a sufficient length of time in one place to determine the economic importance of seasonal fluctuations in the populations of the various naturally occurring pollinators.

While it is possible to infer the effectiveness of naturally occurring pollinating agents by observing fruit-set in orchards where hive bees are absent, to do the reverse would imply a more complete knowledge of insect pollinators than we at present possess. Perhaps the best available evidence relating to the effectiveness of hive bees as pollinators is provided by the literature describing an effect of distance from apiary on yield. Working with apples, Momers (1948a, 1951) not only found the number of bees far less at a distance of 300 yards from the hives (two per acre in one block) than near them, but that the crop in the far part of the orchard was much lower. Reduced yields of apples with increasing distance from the hives has also been reported by Tzyganov (1953) and Glushkov (1958) and a similar situation has also been described for cherries (Davis, 1926; Nevkryta, 1957).

Clearly both hive bees and wild insects are valuable pollinators, the latter especially so where environments favour them. The value of hive bees lies in the fact that they are the only pollinating agent over which the grower has any appreciable control. Their use is justified, not as a means of ensuring a good set, but as a measure of insurance against a bad one.

With so many factors affecting fruit-setting it is in general true to say that no clear-cut distinction can yet be made with regard to successful pollination between the user and non-user of hive bees. The calculation that an insect/blossom ratio of 1 : 30,000 could be effective may help to explain why the orchard into which 50,000 honey bees (one good hive per acre) have been introduced does not necessarily yield better than one in which no hive bees have been used. Firstly, from what we know of their habits, the majority of these bees are selfing and not crossing, and even if cross-pollination is effected to a much greater extent than in the orchard without hive bees, the yields may not appreciably differ, for a high initial-set is normally followed by a high June drop (see Part I, Causes of Fruit Drop in Apples).

Use of Hive Bees in Orchards

Many authors have suggested that where hive bees are used they should be distributed at the minimum rate of one colony per acre (of mature apple trees). The work of Brittain *et al.* (1933) stresses the need to use the area of contiguous orchard as the basis of any calculation, and not the area of the individual plantation it is intended to pollinate. As cherries require a much higher fruit-set than some fruits to produce an economic yield, it would seem logical that a greater quantity of bees should be used for this crop, though insufficient information exists to enable specific recommendations to be made.

From the work of Ribbands (1951) it is obvious that foraging activity and hence pollination is considerably reduced if hive bees are placed away from the crop, especially under unfavourable weather conditions, and Butler (Butler and Simpson, 1954) on the basis of the "wandering bees" hypothesis, suggested that hives should be grouped in the centre of each 15-20 acres of orchard. Apart from the greater ease of management of the colonies provided, these authors point out that by keeping the colonies away from the edges of the orchard the tendency to forage outside them is reduced, and concentrating them in the centre intensifies the competition between bees for forage and thus is likely to create a rise in the percentage of "wandering bees" and consequently increase cross-pollination. Since the radius of a circle of an area of 20 acres is only 176 yards, Butler and Simpson suggest that the consequences of diminished foraging activity in bad weather are minimised.

Free *et al.* (1960) studied the timing of hive placement in relation to blossoming and pollination. They concluded that a greater amount of pollination is likely to follow if hives are not placed in the orchard until flowering has commenced. As it seems probable that the early part of the blossoming period of a variety is particularly conducive to cross-pollination, arrangements should be made well in advance to allow rapid placement of the hives at the critical time.

Artificial Pollination

The inherent uncertainty of the pollination process under orchard conditions, particularly in seasons when poor weather is experienced at blossom time, has led to investigations into the possibility of artificial pollination of fruit trees. In this method pollen is collected in quantity from donor trees and stored under suitable conditions until required, when it is applied to the open blossoms of the receptor trees by hand, by some mechanical means, or by the agency of hive bees.

Pollen is collected by picking large quantities of flowers in the "balloon" stage and spreading them in trays under warm dry conditions until the anthers dehisce. They are then gently rubbed over fine wire sieves to separate the anthers and pollen from the other parts of the flowers. After further screening to clean the pollen from extraneous matter, it is placed in vials for storage until required (Visser, 1955).

Pollen Storage

For apple and pear pollen a relative humdity of 10 per cent is optimal, both higher and lower humidities seriously shortening storage life. A storage temperature of $-20°C$. is effective in retaining viability of fruit tree pollen for periods of 2 to 3 years, and possibly longer (Visser, 1951).

A satisfactory way of applying the pollen to the receptor trees is by hand, using a soft brush which is first dipped into the vial of pollen and then lightly dabbed on the stigmas of the flowers to be pollinated. On an orchard scale this is naturally a time-consuming and costly operation, though it has been used by fruit growers in certain parts of the United States in seasons unfavourable for insect activity. One advantage of hand pollination is that it offers a considerable degree of control over fruit-set, so that by pollinating, for instance, only one flower in every three or four clusters, the possible need for subsequent hand thinning is eliminated.

Hand pollination can be of occasional value also as a means of determining whether or not pollination is a limiting factor. Where the cropping history of an orchard has been bad and it is considered that pollination might be at fault, hand pollination of a number of marked branches or trees scattered throughout the plantation, leaving the flowers exposed to normal pollination as well, might prove worthwhile. If as a result a much higher set was obtained where hand pollination had been carried out, it would be reasonable to assume that insufficient pollination was responsible for low yields in the orchard as a whole (see Table VII). Steps could then be taken to remedy the situation by increasing the pollinator ratio, introducing hives of bees or increasing the number of colonies if these were already in use.

HIVE INSERTS

The use of hive inserts is not new, but only recently have modifications brought encouraging results. In this method previously collected pollen is placed in a device attached to the hive in such a manner that outward going bees become dusted with the pollen, thus carrying it to the trees they visit. Johansen and Degmen (1957) found that bees from hives fitted with suitable inserts were capable of inducing a satisfactory set of fruit on caged apple trees. By using a fluorescent dye, these workers were able to establish that pollen dispensed in this way is carried on to the blossom in the orchard. Townsend *et al* (1958), who tested a modified hive insert in a pear orchard inadequately furnished with pollinators, were able to obtain, over two seasons, large increases in the quantity and uniformity of fruit set.

This means of artificial pollination shows obvious promise, particularly in orchards where cross-pollination is normally inadequate. However, whether there is any advantage to be gained by using inserts in orchards well provided for cross-pollination has yet to be determined. The use of pollen inserts demand not only a readily available source of pollen, but frequent attention to the hives; it is necessary to replenish the pollen hourly to avoid loss of viability.

For application by dust blower or other mechanical means pollen can be mixed with a diluent such as lycopodium powder (Overley and Bullock, 1947), but this method has not proved successful for the simple reason that the stigmas of fruit plants are not well adapted for catching wind-borne pollen grains.

PART III

PLANTING AND POLLINATION

In planning orchards consideration must be given to the optimum spacing and ratio of trees of the pollinator variety relative to those of the main variety. This implies some knowledge of the distance over which cross-pollination by insects is likely to be effective.

Effective Distance for Cross-Pollination

Although precise information on this point is not available, some indication may be gained by the study of orchards containing self-incompatible varieties where insufficient provision for cross-pollination has been made. In such orchards it is often possible to plot the rate at which yield, fruit-set or seed number falls off with increasing distance from the pollinator source.

TABLE VI

Yield* of Bramley's Seedling Apple in Relation to Distance from a
Pollinator

Row	1	2	3	4	7	8	9	10
Distance from west side of plot (feet)	15	30	45	60	105	120	135	150
Period 1. 1925-32 (alternate trees pollinators)	100	82	83	97	96	96	101	107
Period 2. 1936-38 (Pollinators west of row 1 only)	100	79	74	80	62	50	50	60

* Expressed per unit cross-sectional area of trunk. Means of three comparable trees in each row, expressed as a percentage of row 1. The pollinator is 'Worcester Pearmain'.

In a plantation of Bramley's Seedling apple at East Malling (Table VI), yield was reasonably constant across the plot at a time when the distribution and proportion of pollinators was excellent. The removal of these pollinators at a later date left the plot dependent on other trees of the same pollinator variety in an adjacent plot along the west side only. Under these conditions the trees nearest to the source of foreign pollen yielded the highest crop, and the influence of cross-pollination was not apparent at distances of 105 feet and more. Although the trees beyond this distance still yielded half a crop, in certain seasons and situations the effect of distance upon cross-pollination is much more pronounced. An illustration of this is provided by Table VII.

TABLE VII

YIELD OF BRAMLEY'S SEEDLING APPLE IN RELATION TO DISTANCE FROM A
POLLINATOR*

Row	1	2	3	4	5	6
Distance from pollinator (feet) ...	40	80	120	160	200	240
Yield (per cent. of row 1) ...	100	62	51	33	25	20

* From data of Whiffen (1948), pollinator Lord Lambourne.

The apple Bramley's Seedling has the ability to set on occasion at least a partial crop with its own pollen, but sweet cherries are completely self-incompatible and therefore show cross-pollination responses even more clearly. In New York, Tukey (1925) was able to show a sharp reduction in both set and yield where trees were further than one row from a source of foreign pollen, with yield further reduced to 50 per cent at 60 feet. It is nevertheless interesting to note that some foreign pollen was transported to the far side of the block, a distance of 240 feet.

TABLE VIII

FRUIT SET AND YIELD OF WINDSOR SWEET CHERRY IN RELATION TO DISTANCE
FROM A POLLINATOR*

Row	1	2	3	4	5	6	7	8	9	10	11
Distance from pollinator (feet)	20	40	60	80	100	120	140	160	180	200	220
Initial-set natural pollination	43.4	26.4	22.9	16.7	20.6	22.5	23.3	9.8	14.5	—	22.8
Initial-set hand pollination	—	—	35.2	37.2	41.8	46.1	—	—	38.7	—	41.8
Yield from natural pollination per unit cross-sectional area of trunk per cent. row 1 ...	100	75	50	54	45	41	27	37	33	32	28

* From data of Tukey (1925), pollinator Black Tartarian.

While it is obvious from Tables VI-VIII that yield declines with increasing distance from a source of foreign pollen, only an indication is provided of the effective range of cross-pollination in apple orchards.

By using seed content of the fruit as an index of pollination efficiency, however, Williams (1958) was able to show in an orchard of the cider variety Michelin, a self-incompatible diploid, that the useful effect of the pollinating source can extend to approximately 90 feet.

Any estimate of the maximum effective distance between pollinators is of limited practical value, however, because the effective distance will vary with local conditions and climate, since it is dependent on the activity of pollinating insects. It is clear that in designing an orchard it is wise to be cautious and plan to ensure adequate pollination under unfavourable conditions. Thus it seems that the optimal spacing between pollinators is the shortest distance circumstances will allow. In principle each tree should be adjacent to a pollinator.

The desire of many growers to plant, in any one orchard, a single main variety with the minimum number of pollinators, has led to the extensive adoption of the "one-in-nine" system, in which each pollinator is surrounded by eight trees of the main variety. While the effect of distance upon cross-pollination is minimised where this system is used (each tree of the main variety has a pollinator on one side), it is now commonly considered that in many orchards, particularly of apples and pears, higher pollinator ratios would be preferable.

It is also a common experience following inclement blossoming seasons, for old orchards containing many varieties mixed together to yield heavily, while more modern plantings containing fewer varieties do not. This not only suggests that a high pollinator ratio is desirable but that more than one pollinator variety may be advantageous. While there is yet no clear proof that mixed pollens are preferable for cross-pollination purposes, the inclusion of more than one pollinator variety reduces the risk of occasional insufficient cross-pollination due to a small amount of blossom on the pollinator or to seasonal variation in overlap of flowering periods (see Text Fig. 6).

Planting Plans for Pollinators

SINGLE TREE ARRANGEMENTS

A point in favour of single tree arrangements is that the pollinators can be spaced evenly throughout the orchard, thus aiding a uniform distribution of foreign pollen. In addition, although these arrangements may differ widely in pollinator ratios, they provide each tree of the main variety with a pollinator immediately adjacent on at least one side, either directly or diagonally.

The most widely employed system in the past has been the one-in-nine arrangement, in which every third tree in every third row is a pollinator (Text Fig. 4, Plan 1). A valuable property of this system is that any orchard so arranged can be repeatedly thinned without altering either the ratio or distribution of the pollinator trees. Another single tree system is the one-in-four arrangement in which every second tree in every second row is a pollinator (Text Fig. 4, Plan 2a). This has not been extensively used in the past, partly because of the fact that it cannot be thinned without drastically altering the pollinator ratio. However, it is a considerable improvement if pollinator tree positions in adjacent rows are alternated (Text Fig. 4, Plan 2b) as this permits one diagonal thinning without affecting the ratio or distribution of pollinators.

Plan 1

×	×	×	×	×	×	×	×	×	×
×	P	×	×	P	×	×	P	×	×
×	×	×	×	×	×	×	×	×	×
×	×	×	×	×	×	×	×	×	×
×	P	×	×	P	×	×	P	×	×

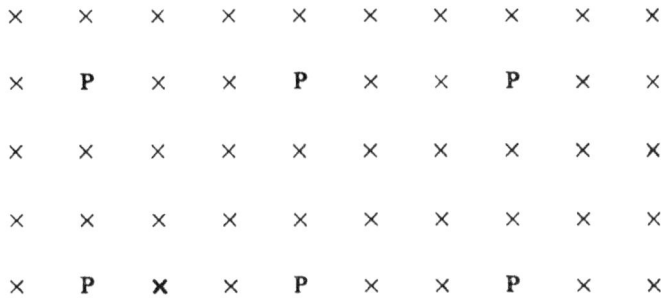

Square plant with 1 in 9 single tree arrangement of pollinators (P)
i.e., every third tree in every third row.

Plan 2

×	×	×	×	×	P	×	×	×	P
×	P	×	P	×	×	×	P	×	×
×	×	×	×	×	P	×	×	×	P
×	P	×	P	×	×	×	P	×	×
×	×	×	×	×	P	×	×	×	P

a   b

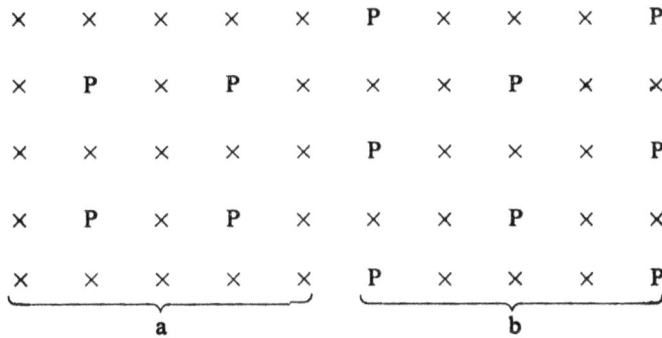

Two arrangements using a 1 : 4 ratio of pollinators
(P) *i.e.*, every second tree in every second row.

Plan 3

P1	×	×	P1	×	×	P1	×	×	P1
P2	×	×	P2	×	×	P2	×	×	P2
P1	×	×	P1	×	×	P1	×	×	P1
P2	×	×	P2	×	×	P2	×	×	P2
P1	×	×	P1	×	×	P1	×	×	P1

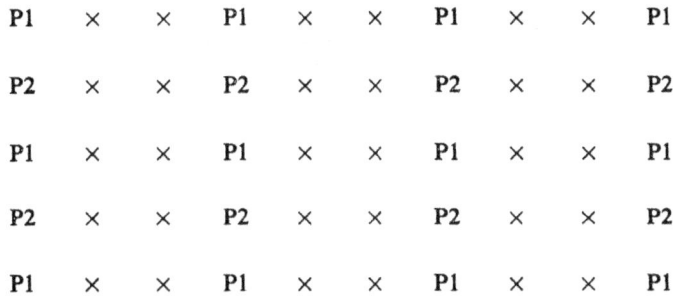

Square plant with 1 in 3 row arrangement of pollinators using
two pollinator varieties (P1 and P2).

Text Fig. 4. Planting Plans, 1, 2 and 3

PLAN 4

```
×      ×      P      ×      ×      P      ×

×      P      ×      ×      P      ×      ×

P      ×      ×      P      ×      ×      P

×      ×      P      ×      ×      P      ×

×      P      ×      ×      P      ×      ×
```

Rectangular plant with 1 in 3 single tree arrangement of pollinators (P) *i.e.*, every third tree in every row.

PLAN 5

```
×      ×      ×      ×      ×      ×      ×

×      ×      ×      ×      ×      ×      ×

×      P      ×      P      ×      P      ×

×      ×      ×      ×      ×      ×      ×

×      ×      ×      ×      ×      ×      ×

×      P      ×      P      ×      P      ×
```

Rectangular plant with 1 in 6 single tree arrangement of pollinators (P) *i.e.*, every third tree in every second row.

All the arrangements illustrated here, with the exception of Plan 2a, can be thinned without altering the proportion or distribution of pollinator trees.

TEXT FIG. 5. PLANTING PLANS 4 and 5

ROW POLLINATOR ARRANGEMENTS

Systems in which the pollinators are arranged in complete rows are now often planted because they give increased ease of management. Row systems generally necessitate a fairly high pollinator ratio, for with bush trees at the common planting distance of 24 feet square, low ratios lead to distances between pollinator and main variety that are too wide for efficient cross-pollination. The introduction of more than one pollinator variety is usually easier than with interspaced arrangements, though this complication may reduce the facility with which a system is thinned.

To provide every tree of the main variety with a pollinator on at least one side, as in single tree systems, every third row must be planted entirely with pollinators; a second pollinator can be added by alternating the rows of pollinator varieties, though from the standpoint of efficient pollination alternating the two varieties down the pollinator rows (Text Fig. 4, Plan 3) is to be preferred.

Alternative row arrangements need not involve a distance between pollinators and main variety greater than twice the planting distance. For instance, four rows of one variety may alternate with up to four rows of another variety.

Rectangular plants lend themselves to the row arrangement of pollinators, but despite the obvious advantages in management that this provides, such arrangements are unlikely to lead to fully efficient cross-pollination (see p. 98). It is therefore desirable to include pollinators in each row and to stagger them as in Text Fig. 5, Plan 4. If such a high pollinator ratio is quite unacceptable on other grounds, then every third tree in every second row giving a ratio of 1 : 5 would ensure that each tree of the main variety suffering from the disadvantage of being in a separate row, would at least be adjacent either directly or diagonally to *two* pollinators (Text Fig. 5, Plan 5).

Sweet cherries pose a special problem in that not only are the planting distances between mature trees very wide, but the arrangement and choice of pollinators is modified by the necessity to take harvesting into account. Because of the problem of picking a large quantity of one variety, it is usual for the orchard to be made up of a series of varieties planted in sequence of ripening order. It is preferable to plant single rows of one variety with a suitable pollinator on at least one side, repeating the sequence of varieties from one side of the field to the other.

Blossoming Period

Flowering begins when the anthers of the first flower to open are visible, and ends when the petals have fallen from the last flower. However, it is usual for a few flowers to open long after the majority, and the duration of the blossoming period is usually expressed as the time interval between the first flower to open and 90 per cent petal fall. While a knowledge of the length of the blossoming period is valuable, it is probably more important to know at what time during this period the tree offers greatest opportunity for cross-pollination; thus many observers record the position of "full" bloom, *i.e.*, the date when 50 per cent of the blossoms are open and 50 per cent are closed. (Beakbane *et al.*, 1935). This is also known as the date of flowering (Irwin, 1931).

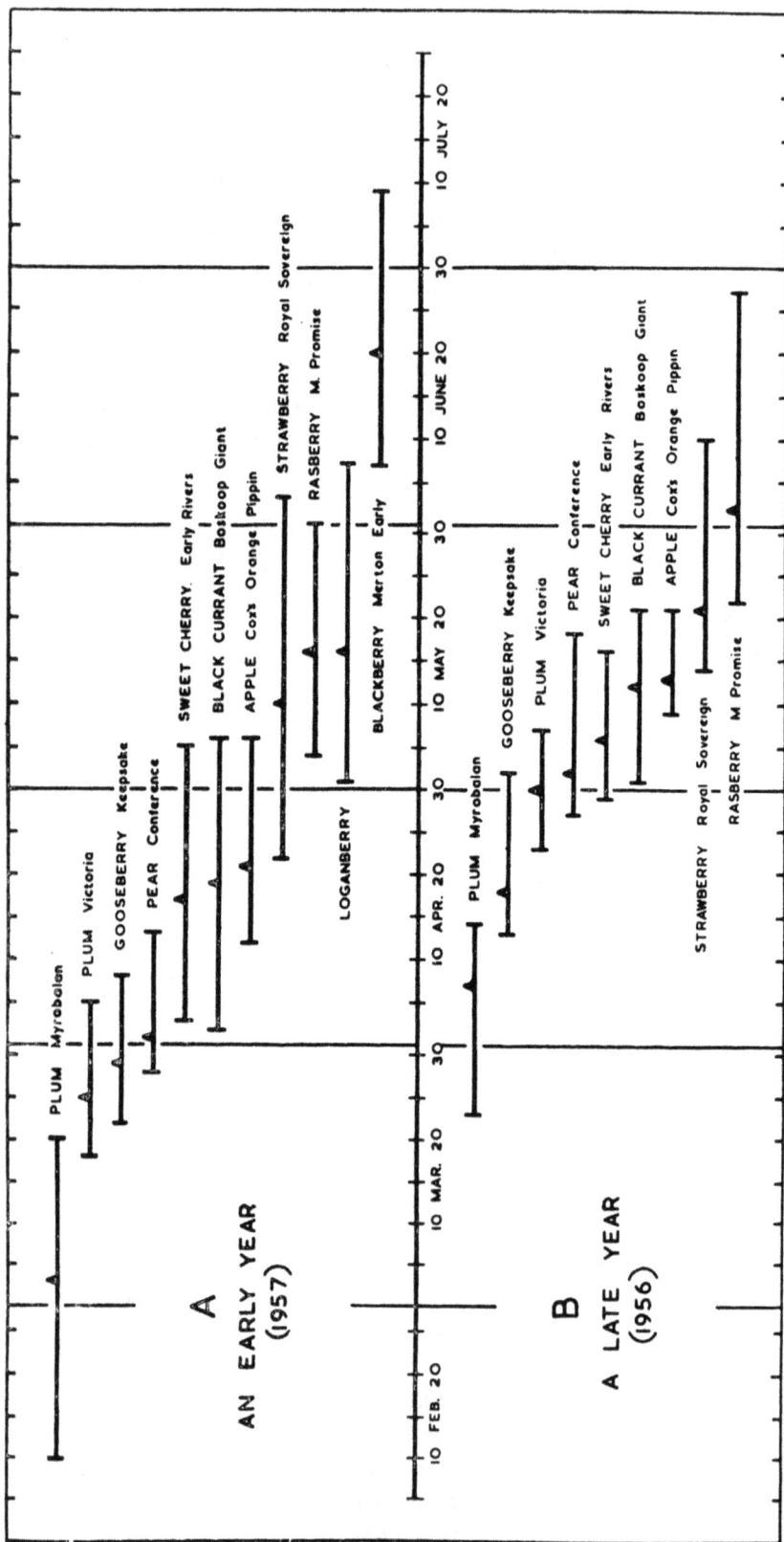

Text Fig. 6. The relationship between the flowering periods of top and soft fruits in a late and an early season (at East Malling).

▲ = Full Bloom.

All varieties are midseason flowering with the exception of Myrobolan plum (very early) and Early Rivers cherry (early).

It has long been known that differences exist between varieties with regard to the length of the blossoming period and the relative position of full bloom (Beakbane *et al.*, 1935) and it is for this reason that care is necessary in the choice of varieties for mutual cross-pollination. The amount of blossom carried by a mature tree is not constant, and in years of reduced blossom production, varieties so affected flower for a shorter time.

The Influence of Temperature

The time of blossoming, however, is predominantly controlled by temperature. Thus Herbst and Weber (1940) in Western Germany determined a threshold temperature of 6°C. (43°F.) for pears, below which flower development within the bud did not proceed. They concluded that the time when a pear variety will flower is determined by the number of hours, calculated from mid-December, during which the temperature exceeds 43°F., this number varying from variety to variety. The variety requiring the least number of degree hours will therefore be the first to flower, although the actual flowering date will vary from year to year. However, such a simple explanation does not seem to be entirely adequate, for varieties tend to vary from year to year in the sequence of flowering which suggests that not all varieties have the same temperature relationships. More recently, Pearce and Preston (1954), working with apples, found that time of flowering was affected not only by temperature during winter and early spring, but also by temperature in the preceding summer, hot weather in June delaying blossoming.

Temperature not only controls the sequence of flowering, but determines the duration of flowering for individual varieties as well as the length of the season between the first and last variety to flower. In some years, once the first variety has come into flower, low temperatures may prevail, with the result that the succession of varieties coming into flower will be very slow and spread over a long period. In other years when temperatures are high the varieties will flower in quick succession and the flowering season will be telescoped and very short.

Varietal Variation

While the variations in flowering time between varieties are often small, in some seasons they cannot be ignored. Because of this year to year variation it is not possible to state definitely that one variety is so much earlier or later than another. Thus, at East Malling, records show that while the average time of flowering of the plum Bryanston Gage is similar to that of Victoria, it can be as much as eight days earlier or four days later. Similarly, Warwickshire Drooper can vary in flowering time between complete coincidence and eleven days earlier than Victoria.

Reference to Table X will show that a similar situation obtains in apples, and that whereas the average variation in time of full bloom between Cox and

Orleans Reinette is such as to suggest that this combination would effectively overlap, Orleans Reinette can flower over a week later. This variety is therefore a less reliable pollinator for Cox than say, Lady Sudeley. It can also be seen that variations in flowering times is greater among early and later flowering varieties than it is among mid-season ones. Comparing the flowering of over 100 apple varieties in Illinois, Crandall (1924) found that no variety proved constant in its flowering position relative to other varieties, and also noted that the greatest variation in this respect occurred among the early varieties.

TABLE X

VARIATION IN THE TIME OF FLOWERING OF APPLES

(Eight years' records at East Malling. Cox's Orange Pippin used as a standard for comparison.)

Variety	Average variation in days	Extreme variation in days
Sunset	− 4.8	−8 to − 2
Egremont Russet	− 4.5	−7 to − 2
Fortune	− 3.6	−8 to 0
James Grieve	− 3.6	−5 to − 2
Lord Lambourne	− 3.0	−5 to + 1
King of the Pippins	− 1.8	−4 to 0
Claygate Pearmain	− 0.8	−4 to + 3
Cox	—	—
Lady Sudeley	+ 0.1	−2 to + 3
Orleans Reinette	+ 2.8	−1 to + 8
Edward VII	+ 5.0	+2 to +11
Crawley Beauty	+13.4	+6 to +23

It is not possible to state the minimum overlap of flowering periods required between two varieties for effective cross-pollination, and it would appear prudent in order to allow for the most unfavourable circumstances likely to be encountered, to cater for maximum overlap. It is thus of some interest to study the distribution of flowers throughout the blossom period.

It can be seen from Text Fig. 7 that whereas the blossoming period extended for 23 days (20 days to 90 per cent petal fall) the period between 50 per cent open blossoms (full bloom) and 50 per cent fallen blossoms extended to 5 days, and further, that the peak blossoming period, i.e., when a minimum of 50 per cent of the flowers were continuously open, extended to only 4 days. Naturally environmental factors can markedly influence the duration of the peak, and Hoblyn (1932) has shown for Bramley's Seedling that 50 per cent full bloom to 50 per cent petal fall can vary between at least 3 and 8 days. Rudloff and Schanderl (1950) recorded the blossom frequency distribution of 24 pear varieties

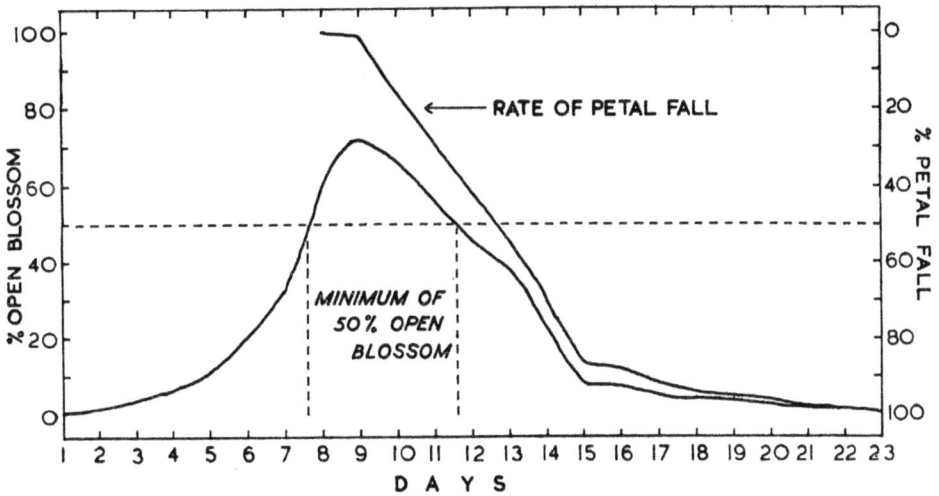

TEXT FIG. 7. BLOSSOM DISTRIBUTION CURVE FOR BRAMLEY'S SEEDLING APPLE.*
* from data of Irwin (1931).

and were able to show distinct differences between varieties with nominally similar blossoming periods. For example, in 1938 the two varieties Comtesse de Paris and Beurré Clairgeau (Text Fig. 8) commenced to flower, reached 50 per cent full bloom and finished flowering on the same dates; however, the duration of the peak blossoming period of Comtesse de Paris was twice the length of that of Beurré Clairgeau.

While such detailed data on blossoming is scant, available information thus suggests that, in order to minimise the effect of year to year variation on cross-pollination, it is wise to plant more than one pollinating variety, and to choose for this purpose varieties whose flowering periods are as similar as possible (see Text Fig. 6).

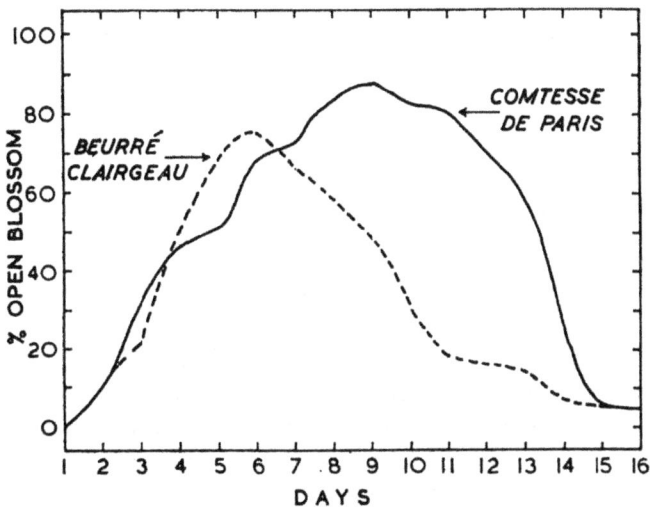

TEXT FIG. 8. BLOSSOM DISTRIBUTION CURVE FOR TWO PEAR VARIETIES WITH COINCIDENT BLOSSOMING PERIODS.*
* From data of Rudloff and Schanderl (1950).

Selection of Pollinators

Much care needs to be exercised in the selection of varieties for effective cross-pollination, and it is the purpose of the appendices to provide information on which a good choice can be made. In general the main points to be considered fall under the following headings.

Compatibility. The choice of a pollinator for sweet cherries and also certain varieties of plums and pears must be carefully checked to see that the intended combinations are reciprocally cross-compatible.

Pollen quality. Varieties with non-dehiscent anthers, pollen of low viability or high sterility make useless or poor pollinators, *e.g.*, Cherry Norman apple and Bristol Cross pear. Where such kinds form the main variety to be pollinated, it is advisable to adopt two pollinator varieties so that these may mutually cross-pollinate each other, though this may be less necessary where the variety chosen as a pollinator is highly self-compatible.

Regularity of flowering. A number of apple and pear varieties have a marked tendency to biennial bearing, and in their "off" years varieties such as Laxton's Superb apple or Emile d'Heyst pear may produce little or no blossom. Such varieties should be avoided as pollinators.

Overlap of flowering period. Top fruit varieties are often classified as early, mid-season, or late flowering, and it is frequently assumed that provided early and late varieties are not planted together, the flowering periods of any two varieties will overlap sufficiently for adequate cross-pollination. This may be so in many years, but as the seasons in which varieties behave abnormally are precisely those in which pollination is most likely to be a limiting factor in crop-production, steps should be taken to guard against this by choosing varieties which flower as nearly simultaneously as possible (see Blossoming). By selecting from the appropriate appendix any two varieties within one flowering group, the best chances of obtaining a consistently satisfactory overlap will be obtained. Varieties from adjacent groups (which in apples differ by only about four days) will provide good overlap of flowering periods in many cases. This will be particularly so in late districts where there is a tendency for flowering periods to be telescoped. In early districts greater care should be taken as in these areas flowering periods are often drawn out.

Varietal efficiency. The view has frequently been held that, irrespective of flowering period, some varieties are better pollinators than others. This is of considerable interest, especially where a pollinator is sought for shy cropping varieties of high potential value, *e.g.*, the pear Doyenné du Comice. The many factors influencing fruit-set make it difficult to assess the relative efficiency of different pollinator varieties. As far as apples and pears are concerned, however, the absence of any clear evidence of differences in setting ability between individual diploid pollens suggests that any two diploids which show suitable characteristics and which flower at the same time should make equally good pollinators.

References

BEAKBANE, A. B., CHAPELOW, H. C. AND GRUBB, N. H. (1935). Periods of blossoming of some tree and soft fruit varieties at East Malling. *Rep. E. Malling Res. Sta. for 1934,* 100-114.

BOHART, G. E. (1952). Pollination by native insects. *Yearb. Agric. U.S. Dep. Agric.,* 107-121.

BRITTAIN, W. H., BLAIR, D. S., GOODERHAM, C. B. AND HERMAN, F. A. (1933). Apple pollination studies in the Annapolis Valley, N.S. Canada 1928-1932. *Bull. Dep. Agric. Can.,* N.S. 162.

BROWN, A. G. (1951). Factors affecting fruit production in plums. *Fruit Yearb.,* 1950, 12-18.

BUTLER, C. G. (1945). The influence of various physical and biological factors of the environment on honeybee activity. An examination of the relationship between activity and nectar concentration and abundance. *J. exp. Biol.,* 21, 5-12.

BUTLER, C. G. AND SIMPSON, J. (1954). Bees as pollinators of fruit and seed crops. *Rep. Rothamst. exp. Sta.* for 1953, 167-175.

CHAMBERS, V. H. (1946). An examination of the pollen loads of *Andrena:* the species that visit fruit trees. *J. anim. Ecol.,* 15, 9-21.

CRANDALL, C. S. (1924). Blooming periods of apples. *Bull. Ill. agr. Exp. Sta.,* 251.

DAVIS, J. J. (1926). Honey bees as an aid in fruit growing. *Amer. Fruit Gr.,* 46, 12.

FABER, H. (1953). Die Bedeutung solitarer Apiden und Fliegen für Bestaubing der Obstbäume nach Untersuchungen im Altenland. I 1951 und II 1952. *Mitt. Obst Vers Anst. Jork.,* 8, 29-39 and 39-40.

FOX WILSON, G. (1926). Contributions from the Wisley laboratory XLVII—Pollination in orchards (VIII) insect visitors to fruit blossom. *J.R. Hort. Soc.,* 58, 125-138.

FOX WILSON, G. (1929). Pollination of hardy fruits: insect visitors to fruit blossoms. *Ann. appl. Biol.,* 16, 602-629.

FOX WILSON, G. (1933). Contributions from the Wisley laboratory LXV—Pollination in orchards (VIII) insect visitors to fruit blossom. *J.R. Hort. Soc.,* 58, 125-138.

FREE, J. B. (1958). Attempts to condition bees to visit selected crops. *Bee World,* 39, 221-230.

FREE, J. B. AND BUTLER, C. G. (1959). Bumblebees. Collins, London.

FREE, J. B., FREE, N. W. AND JAY, S. C. (1960). The effect on foraging of moving honeybee colonies to crops before or after flowering has begun. *J. econ. Ent.,* 53, 69-70.

GLUSHKOV, N. M. (1958). Problems of bee keeping in the U.S.S.R. in relation to pollination. *Bee World,* 39, 81-92.

GRAINGER, H. (1929). *J. Roy. Soc. Arts,* 77, 439.

HERBST, W. AND WEGER, M. (1940). Zur Physiologie des Fruchtens bei den Obstgehölzen. V. Zur Möglichkeit einer Voraussage des Bluhtermines bei den Obstgehölzen, ein Beiträge zum Problem der Temperatursummen. *Forschungsdienst,* 9, 518-523.

HOBLYN, T. N. (1933). Horticultural crop-weather observations in 1932. *J. Min. Agric.,* 40, 146-151.

HOOPER, C. H. (1920). Notes on insect visitors to fruit blossoms. *J. Pomol.,* I, 116-124.

IRWIN, J. O. (1931). Precision records in horticulture. *J. Pomol.,* IX, 149-194.

JOHANSEN, C. A. AND DEGMAN, E. (1957). Progress report on hive inserts for apple pollination. *Proc. Wash. St. hort. Ass.,* 53, 77-81.

LOKEN, A. (1958). Pollination studies in apple orchards of Western Norway. *X Int. Congr. Ent.,* (4), 961-965.

MACDANIELS, L. H. (1931). Further experience with the pollination problem. *Proc. N.Y. St. hort. Soc.,* 76, 32-37.

MASSEE, A. M. (1937). Notes on some interesting mites and insects observed on fruit trees in 1936. *Rep. E. Malling. Res. Sta. for 1936,* 222-228.

MENKE, H. F. (1951). Insect pollination of apples in Washington state. XIV *Int. Bee-keep. Congr.,* Paper 11.

MINDERHOUD, A. (1951). De plaatsvastheid van insecten in verband met de plantenveredeling. *Meded. Dir. Tuinb.,* 14, 61-70.

MOMERS, J. (1948a). Over het aandeel van de honingbijen in de bestuiving van het fruit. *Meded. Dir. Tuinb.,* 11, 252-259.

MOMERS, J. (1948b). De plaatsvastheid der honingbijen. *Meded. Dir Tuinb.,* 11, 529-539.

MOMERS, J. (1951). Honeybees as pollinators of fruit trees. *Bee World,* 32, 41-44.

NEVKRYTA, A. N. (1957). (Distribution of apiaries for pollinating cherries.) *Pchelovodstvo,* 34, 34-38.

OVERLEY, F. L. AND BULLOCK, R. M. (1947). Pollen diluents and application of pollen to fruit trees. *Proc. Amer. Soc. hort. Sci.,* 49, 163-169.

OVERLEY, F. L., O'NEILL, W. S., PAGE, G. M. AND BROWN, C. A. (1946). Experiments with the use of bees for pollination of fruit trees. *Proc. Wash. St. hort. Ass.,* 42, 203-214.

PEARCE, S. C. AND PRESTON, A. P. (1954). Forecasting the time of blossoming in apple trees from temperature records. *Rep. E. Malling Res. Sta. for 1953,* 133-137.

PERCIVAL, M. S. (1947). Pollen collection by *Apis mellifera. New Phytol.,* 46, 142-165.

Preston, A. P. (1949). Observations on apple blossom morphology in relation to visits from honey bees (*Apis mellifera*). *Rep. E. Malling Res. Sta. for* 1948, 64-67.

Rashad, S. E. (1957). Some factors affecting pollen collection by honey-bees and pollen as a limiting factor in brood rearing and honey production. *Kansas State College, Thesis.*

Ribbands, C. R. (1951). The flight range of the honeybee. *J. Anim. Ecol.*, **20**, 220-226.

Roberts, D. (1956). Sugar sprays encourage fertilization by honey bees. *N.Z. J. Agric.*, **93**, 206-211.

Rudloff, C. F. and Schanderl, H. (1950). *Die Befruchtungsbiologie der Obstgewachse und irhe Anwendung in der Praxis.* Eugen Ulmer, Stuttgart.

Rymashevskii, V. I. (1956). (Pollinating activity of bees on the flowers of fruit trees and bushes.) *Pchelovodstvo*, **33**, 51-52.

Sax, K. (1922). Sterility relationships in Maine apple varieties. *Bull. Maine agric. Exp. Sta.*, 307.

Shaw, F. R. (1954). Some observations of the collecting habits of bees. *Amer. Bee J.*, **94**, 422.

Sigh, S. (1950). Behaviour studies of honeybees in gathering nectar and pollen. *Mem. Cornell Agric. Exp. Sta.*, 288.

Stapel, C. (1939). Undersógelser over de ved Frugttraeernes bestovning medvirkende Insekter. *Tidssk. Planteavl.*, **43**, 743-800.

Stephen, W. P. (1958). Pear pollination studies in Oregon. *Bull. Oregon agric. Exp. Sta.*, **43**.

Townsend, G. F., Riddell, R. T. and Smith, M. V. (1958). The use of pollen inserts for tree fruit pollination. *Canad. J. Pl. Sci.*, **38**, 39-44.

Tukey, H. B. (1925). An experience with pollenizers for cherries. *Proc. Amer. Soc. hort. Sci.*, **21**, 69-73.

Tzyganov, S. K. (1953). (Pollination by bees increases the number and improves the quality of fruit.) *Pchelovodstvo*, **30**, 36-38.

Vansell, G. H. (1942). Factors affecting the usefulness of honey bees in pollination. *Circ. U.S. dep. Agric.*, 650.

Vansell, G. H. (1952). Variations in nectar and pollen sources affect bee activity. *Amer. Bee J.*, **92**, 325-326.

Visser, T. (1951). Bloembiologie en krusingstechniek bij appel en peer. *Meded. Dir. Tuinb.*, **14**, 707-726.

Visser, T. (1955). Problemen bij. stuifmeel van fruitgewassen. *Meded. Dir. Tuinb.*, **18**, 856-865.

Wellington, R., Hatton, R. G. and Amos, J. (1921). The "running-off" of blackcurrants. *J. Pomol.*, **2**, 160-198.

Whiffen, H. J. (1948). Bees in the orchard. Facts and figures after using bees for the first time. *Fruit-grower*, **106**, 573-575.

Williams, R. R. (1959). The effective distance of a pollen source in a cider apple orchard. *Rep. Long Ashton Res. Sta. for* 1958, 61-63.

APPENDICES

The aim in the preparation of these appendices has not only been to bring up-to-date many previously existing lists, but to provide the intending planter with a comprehensive source of as much relevant information as is available. A considerable amount of this information has been provided by: The John Innes Institute; East Malling and Long Ashton Research Stations; The National Fruit Trials; The Royal Horticultural Society; The Experimental Horticulture Stations of the National Agricultural Advisory Service; The Kent Farm and Horticultural Institute and Pershore Institute of Horticulture. The following published work has also been freely used.

Anon. (1943). The fertility rules in fruit planting. *John Innes Leafl.*, 4.

Anon. (1957). A revised list of recommended varieties of cider apples. *Rep. Long Ashton Res. Sta. for* 1956, 44-47.

Beakbane, A. B., Chapelow, H. C. and Grubb, N. H. (1935). Periods of blossoming of some tree and soft fruit varieties at East Malling. *Rep. E. Malling Res. Sta. for* 1934, 100-114.

Brown, A. G. (1940). The order and period of blossoming in apple varieties. *J. Pomol.*, **18**, 68-73.

Brown, A. G. (1943). The order and period of blossoming of pear varieties. *J. Pomol.*, **20**, 107-110.

DUGGAN, J. B. (1948). The order and period of blossoming in sweet cherry varieties. *J. hort. Sci.*, **24**, 189-191.

GLENN, E. M. (1957). Blossoming period of some tree fruits at East Malling. *Rep. E. Malling Res. Sta. for* 1956, 89-90.

GRUBB, N. H. (1949). *Cherries.* Lockwood, London.

HOOPER, C. H. (1932). Pollination in relation to cherry orchards. *J. S.-E. agric. Coll., Wye,* **30**, 244-246.

HOOPER, C. H. (1934). Apples, the relative order of flowering of the different varieties and its bearing on cross-fertilization. *J. S.-E. agric. Coll., Wye,* **34**, 210-215.

HOOPER, C. H. (1935). Pears—their pollination, the relative order of flowering of varieties, their cross-fertilization and the insect visitors to the blossoms. *J. S.-E. agric. Coll., Wye,* **36**, 111-118.

HOOPER, C. H. (1936). Plums—notes on their pollination, order of flowering of varieties and insect visitors to their blossoms. *J. S.-E. agric. Coll., Wye,* **38**, 131-140.

JOHNSTON, S. (1951). Essentials of blueberry culture. *Bull. Mich. St. Coll.,* 188.

MERRILL, T. A. (1936). Pollination of the highbush blueberry. *Bull. Mich. agric. Exp. Sta.,* 151.

SCHWARTZE, C. D. AND MYHRE, A. S. (1954). Growing blueberries in the Puget Sound region of Washington. *Circ. Wash. St. Coll.,* 289.

WILLIAMS, R. R. (1953). Pollination requirements for cider apple varieties: progress report for 1952. *Rep. Long Ashton Res. Sta. for* 1952, 44-48.

WILLIAMS, R. R. (1955). Pollination requirements for cider apple varieties: II progress report 1954. *Rep. Long Ashton Res. Sta. for* 1954, 38-46.

APPENDIX I. DESSERT, CULINARY AND CIDER APPLES

All apples show some degree of self-incompatibility; some set no fruit at all when self-pollinated, others can set a fair crop under favourable conditions. Cropping is, however, much more satisfactory and consistent when provision is made for cross-pollination. Although cross-incompatibility has been reported in apples it has not been found among the varieties grown in this country. Apples may be diploid or triploid and the latter are poor pollinators and therefore, in plantations or orchards where these are grown, two diploid varieties should be planted as pollinators to ensure crops from all the varieties unless the pollinating variety chosen happens to be sufficiently self-fertile on its own.

In the following table varieties are divided into seven groups according to the flowering season. In selecting varieties to pollinate each other varieties should be selected from within the same group where possible. Varieties may be selected from the preceding or following group and their flowering periods should overlap sufficiently for cross-pollination to take place with the exception of Group 8. Crawley Beauty (culinary) and Medaille d'Or (cider) usually flower when the flowering stage of all other varieties is over and set an adequate crop on their own.

Whilst in most seasons and districts flowering will follow a regular sequence, variations will occur from year to year and also from district to district. It is known that varieties react differently to winter temperatures and this may well cause some varieties to flower earlier than others in some seasons and later, or at the same time, in others; in the same way variation may occur between varieties growing in the eastern part of England compared with those in the western area in the same year.

Among cider apple varieties, biennial bearing and, therefore, biennial flowering is much more common than it is in culinary and dessert apples. It is doubtful whether cider apple varieties are inherently more biennial in their bearing, but the general standard of nutrition in a cider apple orchard is usually of a much lower standard than is found with dessert and culinary apple orchards and this, in part, tends to make many varieties biennial.

Another point which is essential in planning a cider apple orchard is to know the harvesting time of the varieties, as with grazing of orchards it is an advantage to have all the varieties in any one orchard dropping fruit at more or less the same time. Varieties are, therefore, classified into four groups according to the season of harvest.

In the following tables of dessert, culinary and cider apples, the varieties are divided into eight groups according to flowering season. No dessert or culinary varieties fall into group 7.

Flowering of Dessert and Culinary Apples

1 *Very Early*	2	3	3 *(continued)*	4	5	6	7 None	8 *Very Late*
Aromatic Russet B	Adam's Pearmain B	Arthur Turner	Lobo	Allington Pippin B	Cellini B	Bess Pool	None	Crawley Beauty
Gennet-Moyle T	Baumann's Reinette B	Baldwin TB	Loddington (Stone's)	Annie Elizabeth	Coronation B	Brabant Bellefleur		
Gravenstein T	Beauty of Bath	Belle de Boskoop T	Lord Grosvenor	Ascot B	Frogmore Prolific B	Court Pendu Plat B		
Hume B	Ben's Red B	Belle de Pontoise B	Lord Hindlip	Bow Hill Pippin B	Gascoyne's Scarlet	Edward VII		
Keswick Codlin B	Bietigheimer T	Blenheim Orange TB	Mère de Ménage	Chelmsford Wonder B	King of the Pippins B	Heusgen's Golden Reinette		
Mank's Codlin B	Bismark B	Blue Pearmain B	Merton Pippin	Colonel Vaughan B	Lord Derby	Laxton's Royalty		
Red Astrachan	Carlisle Codlin	Bowden's Seedling	Merton Prolific	Cox's Pomona	Merton Beauty	Reinette Rouge		
	Cheddar Cross	Bramley's Seedling T	Merton Russet	Delicious	Mother (American)	Etoilée		
	Christmas Pearmain B	Brownlee's Russet	Merton Worcester B	Duke of Devonshire	Newton Wonder B			
	Duchess of Oldenburg	Byford Wonder	Miller's Seedling	Dumelow's Seedling (Wellington)	Northern Spy			
	Egremont Russet	Calville Blanche	New Northern Greening	Ellison's Orange	Royal Jubilee			
	George Cave	Charles Ross	Ontario	Golden Delicious	Thurso			
	George Neal	Claygate Pearmain	Peasgood's Nonsuch	Golden Noble	William Crump			
	Golden Spire	Cox's Orange Pippin	Pedro	Gospatrick	Winston			
	Irish Peach	D'Arcy Spice	Potts' Seedling	Herring's Pippin	Woolbrook Pippin B			
	Joyce	Devonshire Quarrenden B	Queen B	Houblon				
	Lawfam	Duchess Favourite	Red Victoria B	Joybells B				
	Laxton's Early Crimson	Early Victoria (Emneth Early) B	Reinette du Canada T	King David				
	Lord Lambourne	Ecklinville	Rival B	Lady Henniker				
	Lord Suffield	Emperor Alexander	Rosemary Russet	Lady Sudeley				
	Maidstone Favourite	Encore	Royal Russet	Lane's Prince Albert				
	Margil	Epicure	St. Cecilia	Laxton's Pearmain				
	McIntosh Red	Exeter Cross	St. Everard	Mannington's Pearmain				
	Melba B	Exquisite	Small's Admirable	Merton Charm				
	Michaelmas Red	Feltham Beauty	Stirling Castle	Mr. Gladstone B				
	Norfolk Beauty	Fortune B	Stonetosh	Monarch				
	Patricia B	George Carpenter B	Sturmer Pippin B	Orleans Reinette				
	Rev. W. Wilks B	Granny Smith T	Sunset	Laxton's Pioneer B				
	Ribston Pippin T	Grenadier	Taunton Cross	Royal Show				
	Ross Nonpareil	Hambling's Seedling	Tom Putt	Sir John Thornycroft				
	St. Edmund's Pippin	Hawthornden	Tydeman's Early Worcester	Sowman's Seedling B				
	Scarlet Pimpernel	Howmead Pearmain	Tydeman's Late Orange	Superb (Laxton's)				
	Striped Beefing	Howgate Wonder	Wagener B	Tydeman's Harvest				
	Warner's King T	James Grieve	Wealthy	Yellow Newtown B				
	Washington T	John Standish	Winter Banana					
	White Transparent	Jonathan	Winter Quarrenden B					
	Withington Fillbasket	Kerry Pippin	Worcester Cross B					
		King's Acre Pippin	Worcester Pearmain					
		Kidd's Orange Red						

B=Known to be bi-ennial or irregular in flowering

T=Triploid

Colour Sports usually flower at the same time as the variety...

FLOWERING OF CIDER APPLES

Flowering Group	Variety		Harvesting Group
2 Early	Dymock Red		2
	Lavignée		3
	Tremlett's Bitter	*	3
3	Sherrington Norman		2
	Red Foxwhelp	*	2
	Cap of Liberty		2
	Improved Foxwhelp	*	2
	Knotted Kernel	*	3
	Court Royal	*T	3
	Upright Médaille d'Or	O	4
	Crimson King	*T	4
	Tardive Forestier		4
4	Langworthy		2
	Reinette d'Obry		4
	Kingston Black Improved		3
	Kingston Black		3
	Frederick	*	3
	Yarlington Mill	*	3
	Sweet Coppin		3
	Bulmer's Norman	*T	2
	Slack Ma Girdle		4
	Reine des Hatives	*	1
	Nehou		1
	Porter's Perfection		3
	Breakwell's Seedling		1
5	Sweet Alford	*	2
	Chisel Jersey		4
	Dabinett	O	4
	Silver Cup		3
	Michelin		3
	Lambrook Pippin		3
	Bedan		3
	White Jersey	*M.S.	3
	Fréquin Audièvre		4
	White Close Pippin		3
	Ashton Brown Jersey	O	4
	Brown's Apple		2
	Fillbarrel		4
	Harry Masters' Jersey		4
	Northwood		3
	White Norman		1
	Fair Maid of Devon		3
6	Cherry Norman	*M.S.	2
	Dove	O	4
	Strawberry Norman	*T	2
7	Brown Snout		3
	Red Jersey		3
	Stoke Red	*	4
	Vilberie		4
8 Very Late	Médaille d'Or	O	4

* Varieties known to need cross-pollination, O = Varieties known to be self-fertile, T = Triploid varieties, M.S. = Male sterile (ineffective as a pollinator). Harvesting periods are divided into groups from (1) early to (4) late.

Appendix II. Dessert, Culinary and Perry Pears

Varieties of pears are less self-compatible than apples and very few fruits are produced from self-pollination. Conference, though sometimes reported as self-fertile, is in fact, self-incompatible, or nearly so. It may set parthenocarpic (seedless) fruits some of which are normal in shape, and this may account for earlier reports that Conference is self-fertile.

Flowering of Dessert and Culinary Pears

1 *Very Early*	2	3	4 *Late*
Brockworth Park	Bellissime d'Hiver	Baronne de Mello	Beurré
Maréchal de la	Beurré Alexandre	Belle-Julie	Bedford M.S.
Cour T	Lucas T	Beurré Brown	Beurré Dumont
	Beurré d'Anjou	Beurré d'Amanlis T	Beurré Hardy
	Beurré Clairgeau	Beurré Fouqueray	Beurré Mortillet
	Beurré Diel T	Beurré Six	Bristol Cross M.S.
	Beurré Giffard	Beurré Superfin	Calebasse Bosc
	Comtesse de Paris	Chalk	Catillac T
	Doyenné d'Eté	Conference	Clapp's Favourite
	Duchesse	Doyenné Boussoch T	Doyenné du Comice
	d'Angoulème	Doyenné George	Early Market
	Easter Beurré	Bouchier	Glou Morceau
	Emile d'Heyst	Dr. Jules Guyot	Gorham
	Fondante Thirriot	Duchesse de	Grégoire's
	Louise-Bonne of	Bordeaux	Bourdillon
	Jersey	Durondeau	Hessle
	Marguerite	Fertility	Laxton's Superb
	Marillat M.S.	Fondante	Laxton's Victor
	Passe Crasanne	d'Automne	Marie Louise
	Précoce de Trévoux	Gansel's Bergamot	Nouveau Poiteau
	Princess	Girogile	Pitmaston
	Roosevelt	Hacon's	Duchess T
	Seckle	Incomparable	Santa Claus
	St. Luke	Herzogin Elsa	Soldat Laboureur
	Uvedale's	Jargonelle T	Winter Nelis
	St. Germain T	Joséphine de Malines	
	Van Mons de Léon	Laxton's Progress	
	Leclerc	Laxton's Satisfaction	
	Vicar of Winkfield	Le Brun	
	Winter Orange	Le Lectier	
		Merton Pride	
		Napoleon	
		Olivier de Serres	
		Packham's Triumph	
		Petite Marguerite	
		Seigneur d'Espéren F	
		Souvenir du Congrès	
		Sucker Pear	
		Thompson's	
		Triomphe de Vienne	
		Verulam	
		Williams' Bon	
		Chrétien	
		Windsor	

T= Triploid   M.S.=Male sterile (ineffective as a pollinator)

Some Synonyms of Pear Names

Synonyms	*True Name*
Beurré d'Esperen	Emile d'Heyst
Conseillier à la Cour	Maréchal de la Cour
Beurré Easter	Easter Beurré
Beurré de Mérode	Doyenné Boussoch
Orange Bergamot	Winter Orange
Beurré Bosc	Calebasse Bosc
Belle Lucrative	Fondante d'Automne

However, parthenocarpy does not always occur and such fruits are often misshapen and of poor market quality. The majority of pear varieties are diploid, though a few are triploid or even tetraploid. Triploid varieties behave in the same way as triploid apples and should have two varieties as pollinators to pollinate both the triploid variety and each other. Two varieties are tetraploids; Improved Fertility, a sport of Fertility is self-fertile; but a sport of Williams' Bon Chrétien known as Double Williams is self-incompatible.

Two incompatibility groups are known in pears. Varieties in these groups are all self- and cross-incompatible, that is, they will neither set fruit with their own pollen nor with the pollen of any variety within the same group. These are:—

Group I	*Group II*
Fondante d'Automne	Beurré d'Amanlis
Laxton's Progress	Conference
Laxton's Superb	
Louise-Bonne of Jersey	
Précoce de Trévoux	
Seckle	
Williams' Bon Chrétien	

Three diploid varieties produce little good pollen and are, therefore, useless as pollinators. These are Bristol Cross, Beurré Bedford and Marguerite Marillat.

The information on perry pears is limited. Preliminary experiments suggest that most varieties are only partially self-fertile in some seasons, therefore provision of facilities for cross pollination is desirable if maximum yields are to be obtained.

In the following tables of dessert, culinary and perry pears, the varieties are divided into 6 groups according to flowering season. There are no dessert pears in groups 5 and 6 nor any perry pear varieties in group 1. Perry pears are also divided into harvesting groups in the same way as cider apples. Where possible, select from the same flowering group when choosing varieties for mutual cross-pollination, though in many cases the flowering period of varieties in adjacent groups will provide sufficient overlap.

FLOWERING OF PERRY PEARS

Flowering Group	Variety	Harvesting Group
2 Early	Hendre Huffcap	2
3	Judge Amphlett	1
	Yellow Huffcap	2
4	Blakeney Red	2
	Brown Bess	4
	Butt	5
	Flakey Bark	3
	Green Horse	4
	Moorcroft	1
	Newbridge	2
	Parsonage	2
	Taynton Squash	1
	Thorn	1
	Winnal's Longdon	3
5	Arlingham Squash	2
	Barnet	2
	Gin	4
	Red Longdon	2
	Oldfield	4
6 Late	Red Pear	3
	Rock	5

APPENDIX III. PLUMS

FLOWERING OF PLUMS

Flowering Group	Compatibility Group A	Group B	Group C	Unclassified
1 Early	Allgrove's Superb Black Prince Grand Duke Jefferson Mallard	Blue Rock Utility	Golden Transparent Goliath	Olympia

2	Admiral
	Black Diamond
	Coe's Golden
	  Drop
	Coe's Violet
	Heron
	Late Orleans
	President
		Angelina Burdett
		Curlew
		Farleigh Damson
			Brahy's
			  Greengage
			Denniston's
			  Superb
			Guthrie's Late
			Langley Bullace
			Monarch
			Ontario
			Prosperity
			Reine-Claude
			  de Bavay
			Warwickshire
			  Drooper
				Mitchelson's

3	Bryanston Gage
	Golden
	  Esperen M.S.
	Kirke's
	Late Orange
	Washington
		Early Favourite
		Early Laxton
		River's Early
		  Prolific
		Goldfinch
		Reine-Claude
		  Violette
			Aylesbury Prune
			Bastard Victoria
			Bountiful
			Brandy Gage
			Czar
			Evesham Wonder
			Laxton's Cropper
			Laxton's Gage
			Laxton's Supreme
			Merryweather
			  Damson
			Pershore
			Purple Pershore
			Severn Cross
			Victoria
				Archduke
				Avon Cross
				Laxton's
				  Abundance
				Swan

4	Count Althann's
	  Gage
	Delicious
	Peach
	Pond's Seedling
	Wyedale
		Belgian Purple
		Cambridge Gage
		Cox's Emperor
		Early Orleans
		Stint
			Blaisdon Red
			Bradley's King
			  Damson
			Early Transparent
			Giant Prune
			Oullins Golden
			  Gage
			Shepherd's
			  Bullace
				Liegel's Apricot
				Teme Cross
				Thames Cross
				Wye Cross

5
 	
Late
	Fellenberg
	Frogmore
	  Damson
	Late Transparent
	Old Greengage
	Red Magnum
	  Bonum
			Belle de Louvain
			Belle de
			  Septembre
			Gisborne's
			Kentish Bush
			Laxton's Blue Tit
			Marjorie's
			  Seedling
			Shropshire
			  Damson
			White Bullace
				Pacific

M.S.=Male sterile (ineffective as a pollinator)

The myrobalan or cherry plums are diploids and are self-compatible. Most other plums, damsons and bullaces grown in this country are hexaploids and may be completely self-compatible, partly self-compatible or completely self-incompatible. All except the completely self-compatible varieties require interplanted pollinators to ensure crops. Cross-incompatibility also occurs, three groups being known.

In a protracted flowering season the time of onset of full bloom from the earliest variety to the latest is about 20 days. In the table which follows, this has been divided into four day periods and the varieties divided accordingly into five flowering groups. When selecting pollinators for varieties which occur in either compatibility Group A or B, the choice is preferably restricted to those varieties whose flowering group is the same as or adjacent to, that of the variety to be cross-pollinated. A pollinator may be selected from any of the three compatibility groups.

GROUPS OF INCOMPATIBLE PLUMS

I	II	III
Jefferson	President	River's Early Prolific
Coe's Golden Drop	Late Orange	Blue Rock
Allgrove's Superb	Old Greengage*	
Coe's Violet Gage	Cambridge Gage	
Crimson Drop		

In Group I: All pollinations fail.

In Group II: Late Orange × President fails both ways.
Late Orange or President pollinated by Cambridge Gage or Old Greengage set a full crop.
Cambridge Gage or Old Greengage pollinated by Late Orange or President set only 2 per cent.

In Group III: Rivers' Early Prolific pollinated by Blue Rock sets a full crop.
Blue Rock pollinated by Rivers' sets a very poor crop.

* Four varieties, perhaps bud sports, are distributed as Old Greengage. They are all in Group II. The differences are mainly in flower and leaf characters.

APPENDIX IV. CHERRIES

Sweet cherries are not only completely incompatible, but a great deal of cross-incompatibility also exists. This means that no variety of sweet cherry will set fruit with its own pollen nor with the pollen of any variety within its own incompatibility group, but will set fruit when pollinated by any variety in another group provided they flower at the same time. The groups are indicated in the following table.

The sour and duke cherries, unlike the diploid sweet cherries, are tetrapoloids. Some are self-fertile, others are self-incompatible and require cross-pollination, but there are no known cases of cross-incompatibility in the sour and duke cherries. Sweet cherries are not suitable pollinators for sour or duke cherries, which however, are capable of pollinating sweet cherries although most of them flower rather too late to be very useful.

The varieties are arranged in groups so that any variety in one of the flowering groups will flower sufficiently close to any other in the same group, or in the group either directly preceding, or immediately following it; for example, the ideal pollinators for Roundel will be found within the same flowering group (group 3), but it could also be pollinated by any variety in flowering groups 2 or 4, provided the variety chosen is not in the same incompatibility group as Roundel. Thus Merton Heart in flowering group 2, Elton Heart in flowering group 3 or Emperor Francis in flowering group 4 will be found satisfactory.

The table is not complete as there are still varieties which require to be tested and placed in their appropriate incompatibility groups. To provide the intending planter with further assistance in preparing a planting plan for a new orchard, or in fitting new varieties into an existing orchard, the sequence of harvesting the various varieties has also been included in the accompanying tables.

FLOWERING OF SWEET CHERRIES

FLOWERING GROUPS

Incompatibility Groups	1 Early	2	3	4	5	6 Late
Universal Donors	Noir de Guben E Goodnestone Black C Nutberry Black C	Mumford's Black B Tartarian E. D	Black Oliver C Bullock's Heart D Merton Glory C	Smoky Dun D	Bigarreau Gaucher F Florence F Smoky Heart F	
I	Early Rivers B	Bedford Prolific C Circassian C Knight's Early Black C	Roundel D Tillington Black D	Black Downton D Ronald's Heart E		
II	Windsor E	Bigarreau de Schrecken C Waterloo D Merton Favourite C	Frogmore Early C Merton Bigarreau C Merton Bounty D	Belle Agathe G	Victoria Black A. D Black Elton D	
III		Bigarreau de Mezel (1) D		Emperor Francis E Napoleon E Ohio Beauty F		
IV	Werder's Early Black B		Merton Premier D	Kent Bigarreau E	West Midlands Bigarreau D	
V		Turkey Heart F			Late Black Bigarreau E	
VI		Merton Heart C	Early Amber C Governor Wood C Elton Heart D			

FLOWERING GROUPS—*continued*

Incompatibility Groups	1 Early	2	3	4	5	6 Late
VII			Bigarreau de Mezel (3) D		Hooker's Black D	Bradbourne Black E Géante de Hedelfingen E
VIII			Peggy Rivers C			
IX	Red Turk E					
X	Ramon Oliva B	Bigarreau Jaboulay B				
XI	Guigne d'Annonay A					
XII					Noble Caroon A F D	
As yet unknown	Rockport Bigarreau B					Cooper's Black F

PICKING SEASON

A Very Early B Early C Early mid-season D Mid-season E Late mid-season F Late G Very late

FLOWERING OF CHERRIES

For sweet cherries	Duke Cherries	Degree of Compatibility	Picking (for sweet cherries)
2	Reine Hortense	S.I.	D
3	May Duke	P.S.C.	C
	Royal Duke	P.S.C.	E
	Empress Eugénie	P.S.C.	C
	Belle de Fraconville		E
4	Belle de Choisy		D
	Archduke	P.S.C.	E
5	Belle de Chatenay	S.I.	G
	Rote Mai		—
6	Ronald's Late Duke	S.C.	G
	Acid Cherries		
3	Olivet	S.I.	D
	Ostheimer Weichsel	P.S.C.	D
4	Kentish Red*	S.I. or S.C.	D
	Wye Morello	S.C.	F
5	Gros-Gobet	S.C.	E
	Montmorency	S.C.	F
	Morello	S.C.	F
	Flemish Red	S.C.	F
6	Coe's Carnation	S.I.	G

S.C. = Self-compatible.
P.S.C. = Partly self-compatible (setting only a light crop when selfed).
S.I. = Self-incompatible.

* Kentish Red exists in two forms under this name, one is self-incompatible and the other is self-compatible.

PICKING SEASON

A Very Early   B Early   C Early mid-season   D Mid-season   E Late mid-season
F Late   G Very late

APPENDIX V. PEACHES

Peaches are self-compatible, however a few varieties such as J. H. Hale produce no good pollen.

A list of flowering sequence is given showing the relationship of other varieties to J. H. Hale.

SEQUENCE OF FLOWERING OF PEACHES IN DAYS

		Opening of First Flowers
Ginard		1st day
Elberta		
Amsden June ...		
J. H. Hale M.S.		2nd day
Duke of York ...		
July Elberta		3rd day
Rochester		
Red Haven ...		4th day
Hale's Early ...		
Early Alexander ...		7th day
Earlyredfre ...		
Peregrine		8th day
Oriole		
Kestrel		9th day
Diana Grace		10th day

M.S. = Male sterile

APPENDIX VI. BLACK CURRANTS

All the commonly cultivated varieties of black currants are self-compatible and no cases of self-incompatible varieties are known. There is evidence to suggest that insects must be present in sufficient quantity for pollination, as cases have been known where very large blocks of black currant bushes have not yielded as heavily as smaller ones, and the bushes nearer to the outside of large blocks have cropped more heavily than bushes in the centre.

There are two varieties under the name of Invincible Giant Prolific. The true variety, which is similar to Goliath, is a good cropper; the rogue variety, on the other hand, although a vigorous grower, sets only a very sparse crop. The unfruitfulness appears to be due to a form of self-incompatibility of an unusual type, in that the pollen tubes quickly reach the ovules but incompatibility is due to failure of fertilisation.

The flowering season, from the first variety to the last, is very short, and there is a considerable overlap of varieties.

The following list places varieties in sequence of flowering:—

Early   Cotswold Cross
         Malvern Cross
         Wellington XXX
         Baldwin
         Westwick Choice
         Westwick Triumph
         Mendip Cross
         Laleham Beauty
         Daniel's September
         Boskoop Giant
         Supreme
         Raven
         French Black (including Seabrook's Black)
         Blacksmith
         Silvergieters Zwarte
         Laxton's Giant
         Goliath (including Edina and Victoria)
         Invincible Giant Prolific
Late   Amos Black

APPENDIX VII. RED CURRANTS

All the cultivated varieties of red currants are self-compatible.

Sequence of flowering from early to late:—

> Fay's Prolific
> Laxton's No. 1
> Red Lake
> Earliest of Fourlands
> Wilson's Long Bunch

APPENDIX VIII. GOOSEBERRIES

All the commonly cultivated varieties are self-compatible. A list divided into groups according to season of flowering is given. In a protracted season there are about four days between the opening of the first flowers of the most forward variety in one group and the first variety of the next group.

FLOWERING SEQUENCE OF GOOSEBERRIES

Early	Early-Midseason	Midseason	Midseason-Late
Cousen's Seedling	Bedford Red	Bedford Yellow	Green Gem
May Duke	Crown Bob	Broom Girl	Careless
Warrington	Green Gage	Early Sulphur	
	Gunner	Lancashire Lad	*Late*
	Ingall's Prolific Red	Leveller	Howard's Lancer
	Keepsake	Speedwell	(Lancer)
	Langley Gage	Thatcher	
	Whinham's Industry	White Lion	
		Whitesmith	

FOOTNOTE.—Information from a number of different sources shows a considerable variation in the flowering of gooseberries from place to place. Whether this is due to faulty nomenclature or to the reaction of varieties to different conditions is not known; the above list should therefore be taken only as a rough guide to the flowering period of varieties.

APPENDIX IX. STRAWBERRIES

Most varieties of strawberries are self-compatible and therefore produce good crops without cross-pollination. The garden strawberry arose from the two octoploid species *Fragaria virginiana* and *F. chiloensis*, this latter species produces plants which are wholly pistillate or wholly staminate. A few varieties have been raised which have inherited the pistillate character; these produce little or no good pollen and will only produce good crops of fruit if cross-pollinated by another variety. This necessitates interplanting a pollinator variety which flowers at the same time as the main variety. Two varieties of this kind, Tardive de Leopold and Oberschlesien gained some popularity, they produce little or no good pollen and are seldom, if ever, grown now.

The sequence of flowering of strawberry varieties appears to vary greatly from one place to another and a list in order of flowering is not given.

FLOWERING OF STRAWBERRIES

Early	Mid-season	Late
Cambridge Favourite	Cambridge Forerunner	Auchincruive Climax
Cambridge Premier	Cambridge Prizewinner	Cambridge Rearguard
Cambridge Profusion	Cambridge Vigour	Merton Princess
Cambridge Rival	Dybdahl	Sir Joseph Paxton
Deutsch Evern	Early Cambridge	Talisman
Huxley	Fenland Wonder	
Madame Lefebvre	Redgauntlet	
Perle de Prague	Royal Sovereign	

APPENDIX X. RASPBERRIES

All the cultivated varieties of Raspberries are self-compatible.

Sequence of flowering from early to late:—

Lloyd George
Malling Promise
Malling Exploit
Malling Jewel
Norfolk Giant

APPENDIX XI. BLACK AND HYBRID BERRIES

The grouping of the following varieties, all of which are self-compatible, is based on blossom records taken at East Malling only.

Early	Mid-Season	Late
King's Acre Berry	Veitchberry	John Innes
Loganberry	Himalaya	
Youngberry	Parsley Leaved Blackberry	
	Bedford Giant	

APPENDIX XII. WALNUTS

Walnuts tend to have a longer blossoming period as they mature, so that even in markedly dichogamous varieties there will be some overlap between the male and female flowering times, and self-pollination will be possible. To ensure ample pollination, however, it is necessary to have at least two varieties growing fairly close together. It is reasonably simple to choose a pollinator for a mid-season variety by using a late-flowering variety, as most of the latter are protandrous and their catkins will mature at the same time as the pistillate flowers of the mid-season variety, but there are very few late-flowering protogynous varieties, which can be used to pollinate themselves and other late-flowering protandrous varieties.

Flowering Periods

Midseason (*protandrous*)

Catkins: early to mid-May
Female flowers: early-mid to end May } usually some overlap.

> Stutton Seedling
> Northdown Clawnut
> Lady Irene

Midseason (*protogynous*)

Female flowers: early-mid May to early June } complete overlap.
Catkins: mid to end May

> Excelsior of Taynton
> Leeds Castle
> Secrett

Late (*protandrous*)

Catkins: mid to end May
Female flowers: end May to mid-end June } usually no overlap.

> Leib Mayette
> Franquette

Appendix XIII. High Bush Blueberries

Most of the information on the pollination requirements for blueberries is available only from the U.S.A. As might be expected from such a large country, where blueberries are grown commercially under very varied conditions and as far apart as the east and west coasts, the information is often conflicting.

Thus the early experiments of Corville (cited by Johnston, 1951) showed that cross-pollination gave better yields in New Jersey. More recent experiments in New Jersey, Massachusetts and North Carolina tend to confirm this result (Johnston, 1951). However, large blocks of a single variety appear to crop satisfactorily in Michigan and Washington (Merrill 1936, Johnston 1951, Schwarze and Myer 1954).

Work has recently started on this problem at Long Ashton Research Station, and the results of the first year, using bushes grown under glass, indicate that there is a large measure of self-fruitfulness though self-pollination results in fewer seeds per fruit. This work needs to be repeated under field conditions. In the meantime, bearing in mind that the need for cross-pollination may well be related to environmental conditions, it is suggested that at least two varieties of blueberries should be planted in each plot or plantation. All varieties appear to overlap extensively in time of flowering.